The Periodic Table

Past, Present, and Future.

Key to cover Periodic Table

Hydrogen is placed centrally (see Chapter 3: 1st Period Problems)

Pale green	Some of the later elements of Period 3 are repeated over the transition metals (see Chapter 9: Group (n) and Group (n+10) Relationships)
Red	The aluminum-iron linkage (see Chapter 9: Group (n) and Group (n+10) Relationships)
Salmon	These are the common knight's move relationship elements (see Chapter 10: Chemical "Knight's Move" Relationships)
Green	These are the common diagonal links (see Chapter 11: Isodiagonality) plus the move of aluminum to its 'correct' place (see Chapter 9: Group (n) and Group (n+10) Relationships)
Pale green	Thorium through uranium are repeated in the transition metals (see Chapter 13: Actinoid and Post-Actinoid Elements)
Grey	The cerium-thorium link (see Chapter 13: Actinoid and Post-Actinoid Elements)
Dark blue	Two most common 'pseudo-elements (see Chapter 14: Pseudo-Elements)
Purple	Two most common 'combo' elements (see Chapter 14: Pseudo-Elements)

This chemistry Periodic Table terminates at Meitnerium. As of the date of writing, no 'real' chemistry of an element beyond Mt has been established. Therefore, no location for subsequent elements can be assigned.

The Periodic Table
Past, Present, and Future.

Geoff Rayner-Canham
Memorial University of Newfoundland, Canada

 World Scientific

NEW JERSEY · LONDON · SINGAPORE · BEIJING · SHANGHAI · HONG KONG · TAIPEI · CHENNAI · TOKYO

Published by

World Scientific Publishing Co. Pte. Ltd.
5 Toh Tuck Link, Singapore 596224
USA office: 27 Warren Street, Suite 401-402, Hackensack, NJ 07601
UK office: 57 Shelton Street, Covent Garden, London WC2H 9HE

British Library Cataloguing-in-Publication Data
A catalogue record for this book is available from the British Library.

THE PERIODIC TABLE
Past, Present, and Future

ISBN 978-981-121-848-4 (hardcover)
ISBN 978-981-121-849-1 (ebook for institutions)
ISBN 978-981-121-850-7 (ebook for individuals)

For any available supplementary material, please visit
https://www.worldscientific.com/worldscibooks/10.1142/11775#t=suppl

Typeset by Stallion Press
Email: enquiries@stallionpress.com

About the Author

Geoff Rayner-Canham, FCIC, FRSC, has published widely on aspects of chemistry education, particularly inorganic chemistry. With Tina Overton, he is coauthor of *Descriptive Inorganic Chemistry*, which is currently in its 6th edition and which has been translated into six other languages. Geoff's main research focus has been on the history of women in science — particularly chemistry. This research has been undertaken jointly with his partner, Marelene Rayner-Canham. In addition to many research papers, they have coauthored six books, the latest being *Pioneering British Women Chemists: Their Lives and Contributions.*

For 20 years, accompanied by some of his students, Geoff has taken chemistry outreach to remote and isolated schools in rural Newfoundland, Labrador, Nunavut, and coastal Quebec. He has also been coauthoring a series of articles with his Inuk student, Chaim Andersen, on Chemistry and Inuit Life & Culture.

With his colleague, Debbie Wheeler, Geoff codeveloped, and has been coteaching, the first online distance chemistry courses offered by Memorial University. One of the courses received a

Canadian award for innovation in distance course technology. Students from as far away as Wuhan, China, and Sydney, Australia, have taken these courses. For his outreach and for excellence in chemistry teaching, Geoff has received the Chemical Institute of Canada, Chemical Education Award; the National Science and Engineering Research Council of Canada PromoScience Award; and a 3M Teaching Fellowship.

Geoff continues to teach and undertake research at the Grenfell Campus, Memorial University, Corner Brook, Newfoundland & Labrador, Canada, where he currently holds the rank of Honorary Research Professor.

Contents

Introduction

"Periodic Properties? That's easy! Properties down a couple of Groups, properties across a sample Period, done!" A not uncommon view. Yet there is a richness of relationships, some obvious, some not, which makes an in-depth look at the chemical elements a rewarding adventure. Ronald Rich eloquently described the lure of periodicity in all its manifestations:

> One of the fascinations of inorganic chemistry is the existence of a wide variety of relationships among the elements and their properties-relationships that show an encouraging degree of order, but a tantalizing variability and novelty. These qualities make the "family of elements" an apt metaphor: while members of a family have much in common, each member also has his[/her] own individual personality.

There have been some 20th century monographs on chemical periodicity. However, to be honest, the old Periodic Table monographs are boring ... no, very boring ... no extremely boring. As are the chapters on the Periodic Table in most textbooks. A litany of dry facts usually emphasizing that everything can be explained in terms of Groups and Periods; that everything is known; that there is only one definitive Periodic Table; and that apart from the genius of Mendeléev, rarely is any other human involvement described.

How incredibly far from the truth in all these factors!

- The Periodic Table is fascinating — as I hope, you, the Reader, will discover.
- Groups and Periods are only one small facet of linkages among the chemical elements.
- There are still avenues of exploration and with many discoveries, new possibilities arise.
- There is no one-fits-all-uses Periodic Table — there are different arrangements to better explain some aspect of element linkages.
- The Periodic Table is a human construct, as can be seen from the names mentioned herein. And in recent times, seven individuals, in particular, have contributed greatly to modern philosophies of the Periodic Table and of the elements therein: Stephen Hawkes, William Jensen, Michael Laing, Pekka Pyykkö, Guillermo Restrepo, R. T. Sanderson, and Eric Scerri. The Reader will see their names (and many others) sprinkled in the text and among the Chapter References.

This book is not a data-filled comprehensive (and boring) compilation. Instead, by looking at some patterns and trends from different perspectives, the Author hopes that the Reader will find this book stimulating and thought-provoking. Without doubt, there are additional interesting and/or curious linkages and patterns of which the Author is unaware. Any Reader spotting an overlooked similarity or pattern is asked to bring it to the attention of the Author at: grcanham@grenfell.mun.ca.

In closing, my Grenfell colleague Chris Frazee, and my partner, Marelene Rayner-Canham, are thanked for reading the entire manuscript (Marelene, many times) in an endeavor to minimize the errors therein.

Geoff Rayner-Canham

Chapter 0

The Periodic Table Exploration Begins!

"The time has come," the Walrus said,
To talk of many things:
Of shoes—and ships—and sealing wax—
Of cabbages—and kings—
And why the sea is boiling hot—
And whether pigs have wings."

Thus spake the Walrus to the Carpenter (Figure 0.1) in *Alice Through the Looking Glass* [1].

Here, in this treatise, Gentle Reader, you will be led through the world of the Periodic Table; a world even **more exciting, more wondrous, more bizarre**, than anything Lewis Carroll could have ever imagined.

Figure 0.1   The Walrus, the Carpenter, and the Little Oysters.

Reference

1. L. Carroll, *More Annotated Alice: Alice's Adventures in Wonderland and Through the Looking Glass and What Alice Found There*, with notes by Martin Gardner; Random House, New York, NY, 220 (1990).

Chapter 1

Isotopes and Nuclear Patterns

In the early decades of modern chemistry, atomic mass (weight) of an element was a major topic for debate and heated dispute. The original Periodic Tables were constructed in terms of order of atomic mass. Any irregularities in order were excused away. With the discovery of atomic number and its use as the foundation of the modern Periodic Table, inorganic chemists seem to have largely ignored patterns in element isotopes. Not only do such patterns explain average atomic mass irregularities, but they reveal some fascinating nuclear chemistry. In addition, the shell model of the nucleus is important in the synthesis of new chemical elements.

In this chapter, the principles of nuclear physics will only be developed to a depth that will aid the understanding of the properties of atoms. For example, the origins of the nuclear strong force, which holds nuclear particles together, is best explained in terms of constituent quarks [1], far beyond the realm of this book. Similarly, the nuclear shell model will be used and applied without delving deeply into its quantum mechanical basis.

Proton–Neutron Ratio

For the lower proton numbers, P, the number of neutrons, N, is approximately matching. With increasing numbers of protons, the numbers of neutrons necessary for nuclear stability increase at a faster rate. For example, the oxygen-16 nucleus has a P:N ratio of 1:1.0, while that of uranium-238 has a P:N ratio of 1:1.6. Figure 1.1 shows a plot of P versus N for stable isotopes [2]. The figure uses the conventional symbol, Z, for the number of protons (from the German, *Zahl*, for "number" [3]). This need for ever-increasing

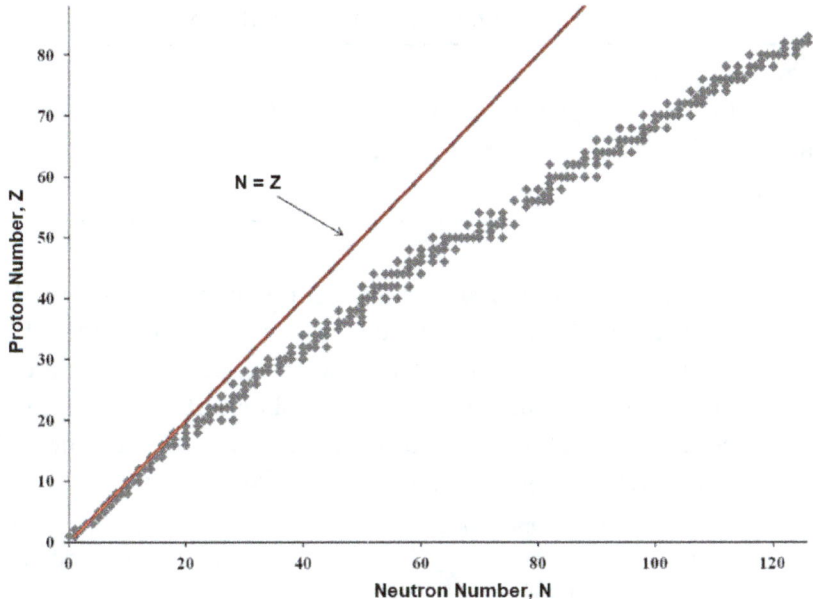

Figure 1.1 Plot of neutrons to protons in stable nuclei (adapted from Ref. [2]).

proportions of neutrons to "stabilize" the nucleus has major implications for superheavy element synthesis as will be shown later in this chapter.

Nuclear Spin Pairing

Different from electron behavior, spin pairing is an important factor for nucleons. In fact, of the 273 stable nuclei, 54% have even numbers of both protons and neutrons (Table 1.1). There is similarly a predominance of even–even nuclei for long-lived radioactive isotopes; those that date back to the origins of the elements [4]. Only four stable nuclei have odd numbers of both protons and neutrons. These stable odd–odd nuclei are hydrogen-2, lithium-6, boron-10, and nitrogen-14 [1]. The only four long-lived odd–odd radioactive isotopes are potassium-40, vanadium-50, lanthanum-138, and lutetium-176.

Table 1.1 Distribution of isotopes

Proton–Neutron Combinations	Even–Even	Odd–Odd	Even–Odd	Odd–Even
Stable	147	4	53	47
Long-lived	20	4	4	4

Spin pairing increases the binding energy; thus, an odd–odd combination has a weaker binding energy than other nuclei, especially even–even. If we look at a series of atoms with the same nucleon (mass) number but differing numbers of protons and neutrons, known as *isobars*, an interesting pattern emerges, known as the Mattauch Isobar Rule:

> The **Mattauch Isobar Rule** *states that: if two adjacent elements in the Periodic Table have isotopes of the same nucleon number, then at least one of the isobars must be a radionuclide (i.e., radioactive).*

This phenomenon is illustrated by the "triplet" isobars, argon-40, potassium-40, and calcium-40, where the argon and calcium isotopes are both stable, while the intervening isobar of potassium is radioactive.

The lack of any stable isotopes of technetium and promethium have always been a notable feature of the Periodic Table. Johnstone *et al.* have used the Mattauch Isobar Rule as a justification of the instability of all technetium isotopes [2]. The neighbors on either side, molybdenum and ruthenium, have six and seven stable isotopes, respectively. These isotopes span the range of "normal" P:N ratios, thus precluding any technetium isotope having a possibility of existence within that range.

The underlying phenomenon was discussed by Suess. He accounted for the instabilities for both technetium and promethium as follows [5]:

> After the filling of the 50- and 82-neutron shell [see discussion below], an upward shift in the β decay energies occurs equivalent to the drop in the binding energy of the last neutron. This shift is somewhat larger, however, for the odd Z than for the odd N nuclei, indicating that the drop is not

equal for paired and for unpaired neutrons.... Thus, for a given I [mass number], the isobars with odd numbers of neutrons become stable at a lower mass number than those with an odd number of protons. This difference is large enough to cause the β-instability of all nuclei with a certain odd number of protons, incidentally those of Z = 43 and 61.

Even Numbers of Nucleons

Elements with even numbers of protons tend to have large numbers of stable isotopes, whereas those with odd numbers of protons tend to have one or, at most, two stable isotopes. For example, cesium (55 protons) has just one stable isotope, whereas barium (56 protons) has seven stable isotopes. The greater stability of even numbers of protons in nuclei can be related to the abundance of elements on Earth. As well as the decrease of abundance with increasing atomic number, we see that elements with odd numbers of protons often have an abundance about one-tenth that of their even-numbered neighbors (see Figure 1.2). This observation is known as the Oddo–Harkins Rule [6]:

*The **Oddo–Harkins Rule** states that an element with an even atomic number is more abundant than either adjacent nucleus with an odd number of protons.*

At the end of the curve, there is a "drop-off." With only radioactive isotopes, the abundances of thorium and uranium have diminished with time. The reduction in abundance over time is also true for other radioactive isotopes, especially potassium-40 [7].

There are two notable exceptions to the Oddo–Harkins Rule. Beryllium would be expected to be much more abundant than it is, while nitrogen would be expected to be significantly less [8]. One might expect beryllium-8 with its 1:1 P:N ratio to be common. However, this nucleus has an extremely short lifetime, splitting into two helium-4 nuclei (helium-4 is "double magic" as we will discuss in the following).

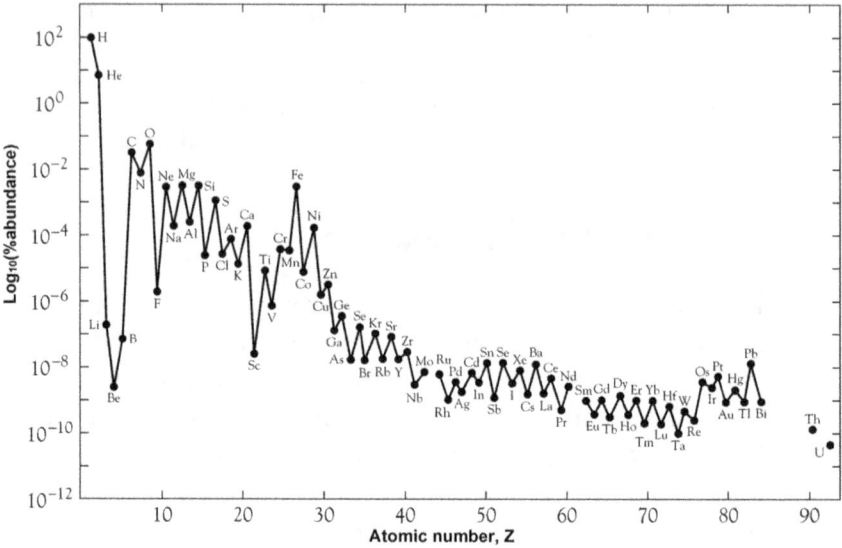

Figure 1.2 A plot of the current abundance of an element against its atomic number.

Nitrogen-14 has an abnormally high abundance, as its formation in stars is part of the CNO nucleosynthesis cycle [9]. The slowest step in the cycle is the proton capture by a nitrogen-14 nucleus. As a result, the cycle often terminates at this step, resulting in an excess of nitrogen atoms compared to the neighboring carbon and oxygen nuclei.

The Cobalt–Nickel and Tellurium–Iodine Atomic Mass Anomalies

Mendeleev organized his Periodic Table according to increasing atomic weight (mass). As Scerri has discussed, there were problems with this rigidity [10]:

The strictest criterion Mendeleev employed was that of the ordering of elements according to increasing atomic weight.... He would occasionally seem to violate this principle, however, in cases where the chemical characteristics of an element seemed to demand it. An example is his placement of tellurium before iodine, as the atomic weight of tellurium has the higher value of the two elements. But while making this reversal, Mendeleev did not just disregard the issue of atomic weight, but rather insisted that the atomic weight of at least one of these elements had to have been determined incorrectly, and that future experiments would eventually reveal an atomic weight ordering in conformity with the placement of tellurium before iodine.

The Cobalt–Nickel Anomaly

In this presumption, Mendeleev was wrong. The measured atomic weights were correct. Generally, as the number of protons in a nucleus increased, so did the number of neutrons in the common isotopes. This accounted for the general correctness of Mendeleev's Periodic Table format. But it is the stable isotopic distribution that could explain the anomalies [11]. The first of these anomalous order pairs was cobalt (58.93) and nickel (58.69). In fact, the Mattauch Isobar Rule applies beautifully as we see in Figure 1.3 for the iron–cobalt–nickel isobars. The only feasible stable isotope of cobalt being cobalt-59.

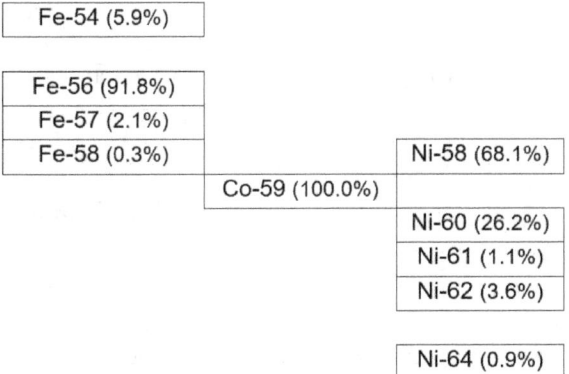

Figure 1.3 Percentages of stable isotopes for the iron–cobalt–nickel sequence.

In fact, the average atomic mass of nickel is less than that of cobalt because of the high proportion of the nickel-58 isotope. The "missing" radioactive even–odd iron-59 isotope decays to the stable odd–even cobalt-59. The cobalt-59 nucleus must represent an energy minimum for the three isobars, as "missing" radioactive even–odd nickel-59 also decays to odd–even cobalt-59 but, in this case, by electron capture.

The Tellurium–Iodine Anomaly

The Mattauch Isobar Rule can be seen for Mendeleev's other example of tellurium (127.6) and iodine (126.9), as shown in Figure 1.4. In this case, iodine-127 is the only possible stable isotope of iodine with a reasonable P:N

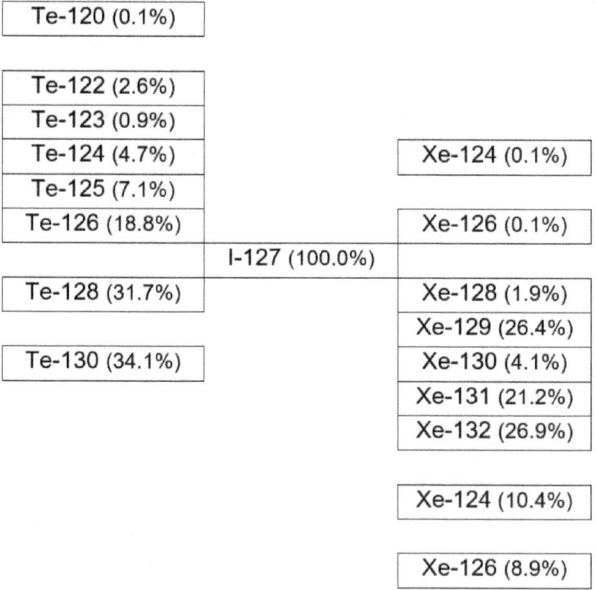

Figure 1.4 Percentages of stable isotopes for the tellurium–iodine–xenon sequence.

ratio. The average atomic mass of tellurium is higher than that of iodine because of the high proportions of the tellurium-128 and tellurium-130 isotopes. The radioactive even–odd tellurium-127 isotope decays to the stable odd–even iodine-127 isotope. Similar to the cobalt-59 situation, the iodine-127 isotope must represent an energy minimum for the three isobars, as radioactive even–odd xenon-127 decays to odd–even iodine-127 by electron capture.

Nuclear Shell Model of the Nucleus

There are two common models of the nucleus, the liquid drop model (which is useful in the context of nuclear fission) and the Meyer–Jensen shell model.

*The **Meyer–Jensen shell model** is a model of the atomic nucleus that invokes quantum principles to explain the nucleon energy levels.*

Such a model is of crucial importance in nuclear spectroscopy. In this chapter, we will look at the use of the Meyer–Jensen model to explain stabilities of isotopes and to show its application in element synthesis. For discussion of the deeper levels of the theory, the Reader must refer elsewhere to the realm of particle physics [12].

According to the nuclear shell model, the principal quantum number, n, for nucleons, like that for electrons, can have values of 1, 2, and so on. However, for nucleons, the angular momentum quantum number, l, is not bound by the principal quantum number (unlike the electron quantum model). That is, for $n = 1$, l can be 1, 2, 3, and so on, resulting in energy levels of 1s, 1p, 1d, and so on. The other two quantum numbers are consistent with the electron case. Thus, there is one 1s quantum state, three 1p quantum states, five 1d quantum states, and so on. Each quantum state can hold two nucleons, one with spin +½ and the other with spin −½. Pauli's exclusion principle also applies to nucleons.

There are two complicating factors for nucleon levels. First, for a polyprotonic or polyneutronic nucleus, as for a polyelectronic system, the energy level degeneracy is lost. That is, 1p will be higher in energy than 1s; 1d higher than 1p; 1f higher than 1d. As a result, 1f is higher in energy than 2s. There is a parallel with electron-level filling, where the 4s orbital can fill before the 3d; the 6s before the 4f; and so on.

Second, the phenomenon of spin–orbit coupling is a major secondary factor for nucleon energies. That is, splitting occurs with a specific level, for example, the 1p level is split into two sublevels, the lower holding a maximum of four nucleons and the upper, two nucleons. To incorporate spin–orbit coupling, a different quantum number, j, is necessary, the total angular momentum quantum number can be the positive values of $(l \pm ½)$. The j value is linked to a matching magnetic orbital quantum number, m_j, where m_j can have values of:

$$(+j); (+j - 1); \ldots 0 \ldots; (-j + 1); (-j)$$

Here we are only interested in the results, not the detailed derivations.

Figure 1.5 shows the energy levels and sublevels up to 70 nucleons. For electrons, an energy "layer" is completed upon filling each np^6. For nucleons, as shown, there is not the same consistency. Instead, it is essentially where there are the larger energy level "gaps." These gaps correspond to filling a total of 2, 8, 20, 50, 82, and subsequently 126, nucleons. On progressing through the remainder of this chapter, the importance of these numbers will become apparent.

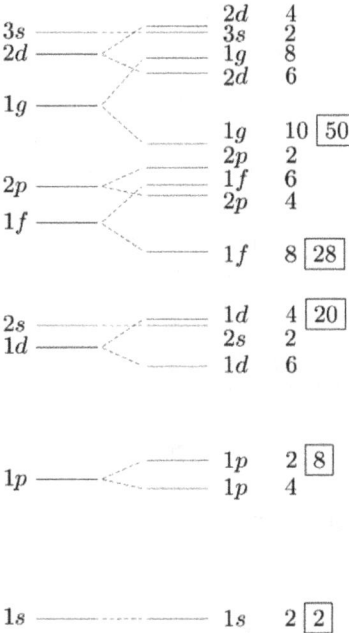

Figure 1.5 Nuclear shell energy levels, the single nucleon to the left, mult-inucleon to the right, up to 70 nucleons.

"Magic Numbers" and Element Isotopes and Isotones

We showed earlier that the nuclei with even numbers of protons had more isotopes than those with odd number of protons. The element with the most stable isotopes is tin, a total of 10. Tin has a "magic number" of protons: 50. These tin isotopes have nucleon numbers of 112, 114, 115, 116, 117, 118, 119, 120, 122, and 124. Not only do the majority of the stable isotopes have even numbers of neutrons, but in terms of abundance, 83.4% of naturally occurring tin has even numbers of neutrons.

Nuclei with the same number of protons and different numbers of neutrons are called isotopes, similarly, nuclei with the same number of neutrons and different numbers of protons are called *isotones*. The largest number of stable isotones are two of the magic numbers: those of $N = 50$ and for $N = 82$. For $N = 50$, the five stable isotones are krypton-86, strontium-88, yttrium-89, zirconium-90, and molybdenum-92. For $N = 82$, the six stable isotones are barium-138, lanthanum-139, cerium-140, praseodymium-141, neodymium-142, and samarium-144.

"Double-Magic" Nuclei

If the possession of a completed quantum level of one nucleon confers additional stability to the nucleus, then we might expect that nuclei with filled levels for both nucleons — so-called doubly magic nuclei — would be even more favored. This is indeed the case. In particular, helium-4 with $1s^2$ configurations of both protons and neutrons is the second most common isotope in the universe, and the helium-4 nucleus (the α-particle) is ejected in many nuclear reactions. Similarly, it is the next doubly completed nucleus, oxygen-16 (8P, 8N), which makes up 99.8% of oxygen on this planet. As we saw in Figure 1.1, the number of neutrons increases more rapidly than that of protons. Thus, the doubly stable isotope is lead-208 (82P, 126N). This is the most massive stable isotope of lead and the most common in nature.

Some doubly magic nuclei can be created whose P:N ratio is far from the usual range of values and physicists have sometimes set out specifically to synthesize the missing one of the pair. Tin provides one such example. Tin-132 (50P, 82N) is a long-known radioactive isotope with a half-life of 40 s. Then in 1994, highly neutron-deficient tin-100 (50P, 50N) was synthesized [13]. It has a short half-life of 1 s, but considering it has such an abnormal P:N ratio, it is surprising that it is that long-lasting.

Calcium is even more evidence of the doubly magic pair phenomenon. The only stable isotope of calcium is the significantly neutron-deficient calcium-40 (20P, 20N). The other doubly magic calcium isotope is the neutron-rich calcium 48 (20P, 28N), which has an exceptionally long half-life of almost 10^{20} years. A "doubly magic" trio is that of nickel [14]. Nickel isotopes are known for nickel-48 (28P, 20N); nickel-56 (28P, 28N); and nickel-78 (28P, 50N).

More "Magic Numbers"?

In recent years, there has come the possibility of synthesizing nuclei very far from the normal P:N ratio. In such circumstances, there seem to be additional closed shell values, resulting in new "magic numbers." There is a specific interest in exotic "doubly magic" atoms. A definition of a doubly closed shell nucleus is a spherical shape and abnormally high energy needed to raise a nucleon to an excited state. One exceptionally neutron-rich nuclei that fit the criteria is oxygen-24 with a ratio of 1:2.0, indicating that 16 neutrons may provide a "magic number" in this circumstance [15]. Similarly, calcium-54 shows similar characteristics, suggesting another "magic number" of 34 under such a high ratio of 1:1.7 [16].

Limits of Stability

In the universe, there are only 80 stable elements (Figure 1.6). For these elements, one or more isotopes do not undergo spontaneous radioactive decay. No stable isotopes occur for any element after lead, and two elements in the earlier part of the table, technetium and promethium (both mentioned earlier) exist only as radioactive isotopes.

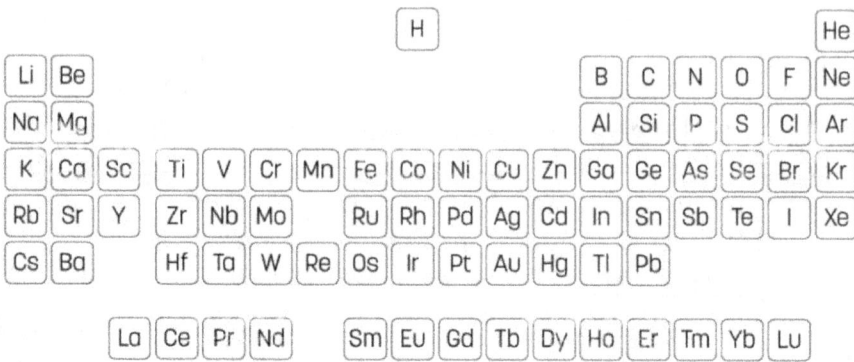

Figure 1.6 Periodic Table showing elements with one or more stable isotope.

Traditionally, bismuth, or more correctly bismuth-209, was considered the last stable isotope. However, as early as 1949, it was predicted theoretically that the isotope could not be stable. It was not until 2003 that the radioactive decay of the "stable" isotope of bismuth was observed [17], and its half-life has now been determined as 1.9×10^{19} years. Beyond the "magic number" of 126 protons of lead, the number of positive charges in the nucleus becomes too large to maintain infinite nuclear stability, and the repulsive forces prevail.

Two postlead elements for which only radioactive isotopes exist, uranium and thorium, are found quite abundantly on Earth because the half-lives of some of their isotopes — 10^8 to 10^9 years — are almost as great as the age of Earth itself.

Synthesis of New Elements

A goal of both chemists and physicists has been the synthesis of atoms of new chemical elements. Such atoms

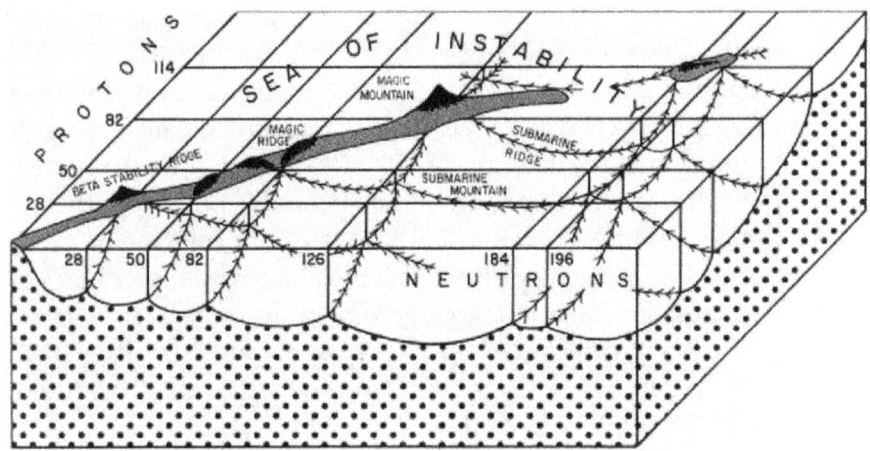

Figure 1.7 The original geographic plot by Seaborg of isotope stability (from Ref. [19]).

generally have very short half-lives. In fact, in order to claim synthesis of a new element, the isotope must have a half-life longer than 10^{-14} seconds, thus excluding "quasi-atoms," briefly existing species formed during nuclear collisions [18]. Seaborg designed a plot of the stability of elements using a geographical analogy of islands in a sea [19]. The goal was to reach the "island of stability" (Figure 1.7). Many variations and updates have appeared since this first one, all based upon the same theme.

To accomplish such syntheses, a target of a high atomic number element is bombarded with atoms of a neutron-rich element whose combined atomic number is that, or greater than that, of the desired element. Up to p = 112, the more common route for the synthesis of postactinoid elements was the use of "doubly magic" lead-208 or "singly magic" bismuth-209 as targets and stable neutron-rich nickel-64 or zinc-70 as projectiles [20]. However, to synthesize the "doubly magic" hessium-270, californium-248 was bombarded with magnesium-26 [21].

Beyond $P = 112$, a different route was chosen. This method involved taking atoms of one of the later actinoid elements and bombarding them with calcium-48 as a projectile. About 0.2% of natural calcium is neutron-rich "doubly magic" calcium-48 (20P, 28N). With a neutron–proton ratio of 1:1.4, calcium-48 has been the key to synthesizing many new elements. Using calcium-48 nuclei as projectiles, nuclear physicists have claimed the synthesis of isotopes of element 114 (Fl) from plutonium-244; element 115 (Mc) from americium-243; element 116 (Lv) from curium-248; element 117 (Ts) from berkelium-249; and element 118 (Og) from californium-249.

Now the aim is to make the first elements of the next period [22]. *"Quo Vadis?"* stated Karol, as he reviewed the nuclear challenges [23]. There are no long-lived target isotopes with even higher atomic number, while the most probable higher atomic number projectile would be titanium-50. Titanium-50 (abundance 5.2%) has the same magic number of neutrons as calcium-48 with two more protons. Therefore, impacting californium-249 should enable the synthesis of an isotope of element 120. Unfortunately, with a P:N ratio of only 1:1.27, it is less likely that long half-life atoms of the desired atomic number would be produced. Similarly, an isotope of element 119 might be expected from the impact of titanium-50 on berkelium-249. But to raise the probability of success, the focus is on the even-proton-numbered element 120.

Island of Stability

Much of the interest in the synthesis of new elements is the belief that, approaching the next set of magic numbers of protons and neutrons, the trend for ever-shorter half-lives will be reversed. As mentioned earlier, this P-N region has been called the island of stability. There have been predictions that the island is centered around P = 114 and N = 184 and encompass combinations of proton and neutron values around that. Many proposals have been made about the island's precise location. The major difficulty is in the synthesis of nuclei with high enough numbers of neutrons to confer longer half-lives. Some of the more optimistic calculations have suggested half-lives in days, years, and even millions of years [24].

It has been claimed that other islands of stability may exist. The first of these would be around P = 126, N = 216 or 228; and the second near P = 164, N = 308 or 318. Only time and nuclear experimentation will tell whether these nuclei are forever beyond the limits of synthesis. The key to reaching any of the islands is, of course, adding enough neutrons to generate a high enough P:N ratio [25].

Commentary

Chemists so often overlook the fascinating world of nuclear structure. So much can be explained. In fact, knowing about P:N ratios and magic numbers is key to understanding the difficulties of synthesizing new elements.

References

1. A. Millevolte, "Nuclear Stability and Nucleon-Nucleon Interactions in Introductory and General Chemistry," *J. Chem. Educ.* **87**, 392–395 (2010).
2. E. V. Johnstone *et al.*, "Technetium: The First Radioelement in the Periodic Table," *J. Chem. Educ.* **94**, 320–326 (2017).

3. W. B. Jensen, "The Origins of the Symbols *A* and *Z* for Atomic Weight and Number," *J. Chem. Educ.* **82**(12), 1764 (2005).

4. "Even and odd atomic nuclei," Wikipedia, https://en.wikipedia.org/wiki/Even_and_odd_atomic_nuclei, accessed 28 June 2019.

5. H. E. Suess, "Magic Numbers and the Missing Elements Technetium and Promethium," *Phys. Revs.* **81**(6), 1071–1072 (1951).

6. A. M. Nikanorov, "Oddo-Harkins Evenness Rule as an Indication of the Abundances of Chemical Elements in the Earth's Hydrosphere and Estimations of the Nature of Cosmic Bodies," *Geochem. Intnl.* **54**(5), 464–469 (2016).

7. T. P. Kohman, "Chronology of Nucleosynthesis and Extinct Natural Radioactivity," *J. Chem. Educ.* **38**(2), 73–82.

8. S. I. Dutch, "Periodic Tables of Elemental Abundance," *J. Chem. Educ.* **76**, 356–358 (1999).

9. J. Audouze, J. W. Truran, and B. A. Zimmerman, "Hot CNO-Ne Cycle Hydrogen burning. I. Thermonuclear Evolution at Constant Temperature and Density," *Astrophys. J.* **184**, 493–516 (1973).

10. E. R. Scerri, *The Periodic Table: Its Story and Its Significance*, Oxford University Press, Oxford, 125–126 (2007).

11. F. H. Firsching, "Anomalies in the Periodic Table," *J. Chem. Educ.* **58**(6), 478–479 (1981).

12. R. D. Lawson, *Theory of the Nuclear Shell Model*, Oxford University Press, Oxford (1980).

13. M. Simm and D. Clery, "Physicists Find a Double Dose of Magic," *Science* **264**, 777 (1994).

14. G. W. Rayner-Canham, "Nickel-48: Double Magic," *Educ. Chem.* **38**, 46–48 (2001).

15. R. V. F. Janssens, "Unexpectedly Doubly-Magic Nucleus," *Nature* **459**, 1069–1070 (2009).

16. D. Steppenbeck *et al.*, "Evidence for a New Nuclear 'Magic Number' from the Level Structure of ^{54}Ca," *Nature* **502**, 207–210 (2013).

17. P. de Marcillac, "Experimental Detection of α-Particles from the Radioactive Decay of Natural Bismuth," *Nature* **422**, 876–878 (2003).

18. S. Hofmann *et al.*, "On the Discovery of New Elements (IUPAC/IUPAP Provisional Report)," *Pure Appl. Chem.* **90**(11), 1773–1832 (2018).

19. G. T. Seaborg, "Prospects for Further Considerable Extension of the Periodic Table," *J. Chem. Educ.* **46**(10), 626–634 (1969).

20. Y. T. Oganessian *et al.*, "Synthesis of a New Element with Atomic Number $Z = 117$," *Phys. Rev. Lett.* **104**(14), 142502-1/4 (2010).

21. J. Dvorak, "Doubly Magic $^{270}_{108}\text{Hs}_{162}$," *Phys. Rev. Lett.* **97**, 242501–242504 (2006).

22. V. Zagrebaev *et al.*, "Future of Superheavy Element Research: Which Nuclei Could Be Synthesized within the Next Few Years?" *J. Phys. Conf. Ser.* **420**, 012001 (2013).

23. P. J. Karol, "Heavy, Superheavy ... Quo Vadis?" in E. Scerri and G. Restrepo (eds.), *Mendeleev to Oganesson: A Multidisciplinary Perspective on the Periodic Table*, Oxford University Press, Oxford, 8–42 (2018).

24. Y. T. Oganessian and K. P. Rykaczewski, "A Beachhead on the Island of Stability," *Phys. Today* **68**, 32 (01 August 2015).

25. P. J. Karol, "The Mendeleev-Seaborg Periodic Table: Through $Z = 1138$ and Beyond," *J. Chem. Educ.* **79**, 60–63 (2002).

Chapter 2

Selected Trends in Atomic Properties

This book cannot be a comprehensive compilation of all the periodic properties. And it is certainly not intended to be a turgid, endless, boring, collection of tabulated data and graphical plots. In this chapter, there will be a focus just upon four major parameters: electronegativity, ionization energy, electron affinity, and relativistic factors. But first ...

What is a chemical element? This question may seem self-evident, but it is not. In Chapter 1, a chemical element was defined, by inference, as an atom with a specific number of protons. Now, entering the world of chemistry, things become murkier. Many electrons have been put to work to produce erudite articles on the subject [1–3].

An atom clearly is a chemical element. It is a minuscule fragment of matter with no color, and no sense of whether it is supposed to be a metal or nonmetal; or have any other bulk properties. It does have an electron configuration, an ionization energy, and an electron affinity. However, to define its electronegativity (discussed in the next section), comparison must be undertaken using a pair of dissimilar atoms.

But most chemists deal with real, visible materials, that is, "elements and compounds," rather than "atoms and molecules." Thus, the question arises, if the word "element" is used to describe the identity of an atom, can it also be used to do double duty for a bulk collection of atoms [4]? And, indeed, at what size does a cluster of atoms begin to exhibit bulk properties [5]?

The proposal has been made to use the term "elementary substance" for a bulk substance that does not undergo chemical decomposition into other substances [6]. There are two challenges with adoption of this new terminology: first, the cumbersomeness of a two-word term; and second, the acceptance among, not just chemists, but also the wider community at large. As much as this author, in general, espouses precise and correct terminology wherever feasible, here, the word "element" will continue to do "double duty."

Electronegativity

It may seem odd that the first parameter in this chapter is not a simple measurable property of an atom. However, the concept of electronegativity underlies much of our interpretation of chemical properties and behavior. Jensen has traced the origins of the electronegativity concept back to the early 1800s, with Berzelius naming the concept "electronegativity" in 1811 [7]. In fact, it was in the late 19th and first half of the 20th centuries when the concept became refined into its modern form [8].

Electronegativity as a Fundamental Property

Leach has produced a comprehensive study of the various parameters that have been used to obtain numerical values for electronegativity. He argued that even though electronegativity is not a single parameter of an atom's properties, it is a fundamental property itself. Among his conclusions were the following (edited) statements [9]:

- Electronegativity is an extremely successful but ill-defined heuristic concept for the description of central properties of entities in the dappled chemical world.

- Electronegativity is a theoretical construction than a natural property. That is, it is cannot be measured directly.
- Electronegativity's main applications (descriptions of polarities and bonding modes in substances, depiction of oxidation numbers, explanation of reaction mechanisms and acidity, etc.) are qualitative. In Leach's view, the mathematization of electronegativity is excessive and tends to lead to *apparent scientificity*.
- Electronegativity is a dimensionless number that — like other measures in the applied sciences — has a complex referential background. It is conceptually rooted in the realm of chemical reactivity on the one hand, but it is supplied by the physics of isolated particles on the other.

Electronegativity Scales

Though electronegativity is commonly associated with Pauling and his scale [10], there are many other scales, including the widely used Allred–Rochow [11]. All numerical ranges share a common definition:

> *Electronegativity* is a chemical property that describes the tendency of an atom to attract a shared pair of electrons (or electron density) toward itself.

If electronegativity is such a vague concept with disparate definitions, why is it so important, and why has it not been replaced by a clearly quantifiable atomic parameter? It was Rodebush in 1924 who provided an elegant answer [12]:

> I had hoped that we might be able to substitute electron affinity or ionizing potential for the wretched term electronegativity, but these quantities are measured for the gaseous state and our ordinary chemical properties are concerned with the condensed phases.

Even in the 21st century, the nature of electronegativity is a continuing topic of discourse and debate [13, 14].

The Sanderson Electronegativity Scale

In addition to the well-known scales, in an oft-overlooked series of articles, Sanderson developed and applied his own electronegativity scale [15]. Though his value system did not gain acceptance, the plot that he generated provided a useful qualitative overview of comparative electronegativities (Figure 2.1).

Sanderson indicated in his plot that the "secondary" linkages progressed from the 3rd Period to the 4th Period. Of particular note in the context of Chapter 9, he highlighted in the article the closer resemblance in electronegativity (and electron configuration) of aluminum with the Group 3 elements, and of silicon with the Group 4 elements. The plot also showed the "saw-tooth" pattern in electronegativities that are observed when descending groups in the later main group elements.

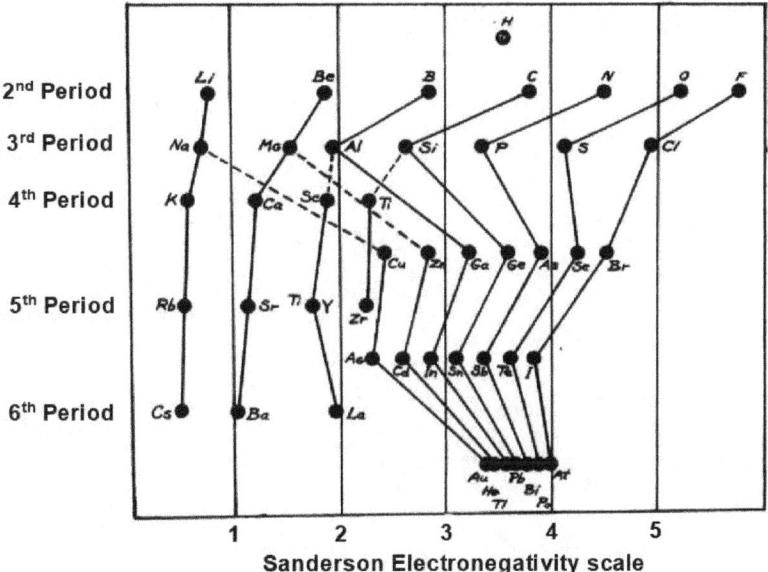

Figure 2.1 Patterns in the Sanderson electronegativities of the elements (adapted from Ref. [15]).

The van Arkel–Ketelaar Bond Triangle

One of the many applications of the electronegativity concept is that of the bond triangle [16]. Devised by van Arkel, then improved upon by Ketelaar, the triangle categorizes elements, alloys, and compounds according to their degrees of ionic, covalent, and metallic character [17]. The triangle was originally more conceptual than detailed. However, using electronegativity values, Jensen semiquantified the diagram as shown in Figure 2.2 [18].

Sproul and his coresearchers have refined and updated the bond triangle, both from a practical and theoretical perspective [19]. The triangle has become popular as a means of assigning bonding in newly synthesized solid-state compounds, such as Y_3AlC [20] and the $AgXY_2$ family where X is a Group 13 element, and Y is a Group 16 element [21]. In addition, Sproul has attempted to revive the triangle in the context of teaching bond type at university level [22].

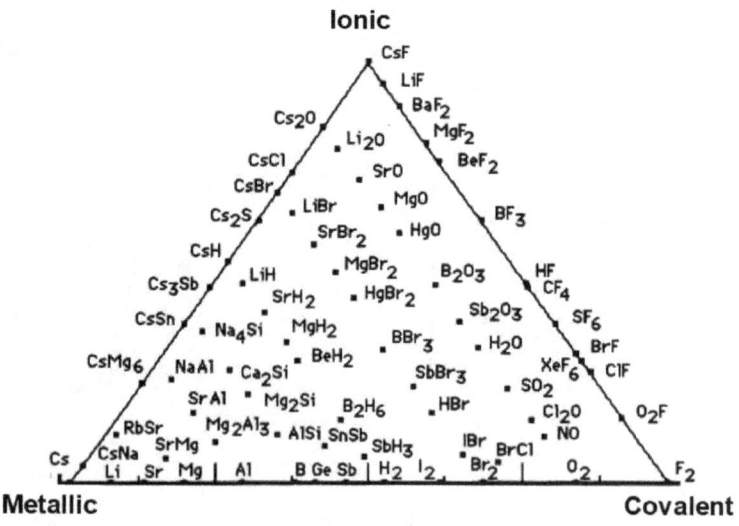

Figure 2.2 A semiquantitative bond triangle (adapted from Ref. [15]).

In Britain, an advanced high school qualification, called the Pre-U has been devised by the University of Cambridge International Examinations. A key part of the chemistry Pre-U is devoted to the van Arkel–Ketelaar Triangle as a basis for the bonding discussion [23].

Oxidation State

Another rather nebulous — but very useful — property is oxidation state [24]. To use the deductive approach as will be shown shortly, the value is determined upon the basis of comparative electronegativities.

According to Jensen, the first use of the term was by Talbot and Blanchard in 1907 [25]. Actually, there are two terms: "oxidation state" and "oxidation number." In 1990, the International Union of Pure and Applied Chemistry (IUPAC) provided a set of recommended lengthy rules by which the "oxidation state" of any element could be calculated (given in Arabic numerals). IUPAC reserved the term "oxidation number" for the central atom in a coordination compound (given in Roman numerals) [26]. This latter usage corresponded to the use of Roman numerals in the Stock system of inorganic nomenclature.

The 1990 IUPAC definition of oxidation state was critiqued by Loock, who pointed out the significant shortcoming that the IUPAC method would only give an average oxidation state if there were two atoms in a compound in dissimilar environments. Instead, he proposed a succinct definition based on Pauling's use of the term [27]:

> The **oxidation state** of an atom in a compound is given by the hypothetical charge of the corresponding atom ion that is obtained by heterolytically cleaving its bonds such that the atom with the higher electronegativity in a bond is allocated all electrons in this bond. Bonds between like atoms (having the same formal charge) are cleaved homolytically.

Jensen strongly supported the Pauling/Loock definition, adding a summary of the difference in the approaches [28]:

> *...the memorized IUPAC rules ...are ultimately traceable to an attempt to assign oxidation values based solely on the use of a species' compositional formula, whereas the Pauling[/Loock] approach requires instead a knowledge of the species' electronic bonding topology as represented by a Lewis diagram.*

A variation of the Pauling/Loock approach — the "exploded structure method" — was devised by Kauffman [29], though its shortcomings were described by Woolf [30]. IUPAC subsequently reversed their view, changing from a mechanical rule-based approach (still widely used in textbooks) to the electronegativity–Lewis structure approach of Pauling/Loock. However, compared with Loock's definition, the IUPAC definition seems technical and obtuse [31]:

> *The oxidation state of an atom is the change of this atom after ionic approximation of its heteronuclear bonds. Bonds between atoms of the same element are not replaced by ionic ones: they are always divided equally.*

In many cases, both the algebraic and the electronegativity–Lewis structure approach give the same result. For example, in the sulfate ion, both methods assign an oxidation state of +6 for sulfur. However, very different results are obtained where there are two (or more) atoms of the same element in different environments.

An example is provided by the thiosulfate ion, $S_2O_3^{2-}$, with a peripheral and a central sulfur atom. When this ion decomposes in acid, the fates of the two sulfur atoms are quite different, indicating that they have come from very different environments and oxidation states in the thiosulfate ion itself. However, the algebraic calculation provides an average oxidation state of +2 for each sulfur atom. The

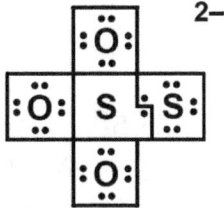

Figure 2.3 Electron assignment for the thiosulfate ion to use for electronegativity determination by the Pauling/Loock deductive method.

Lewis structure of the ion (Figure 2.3) confirms the experimental finding of two very different electron environments for the sulfur atoms. Utilizing the Pauling/Loock electronegativity–Lewis structure approach, the central sulfur atom is assigned an oxidation state of +5 while the peripheral one has a resulting oxidation state of −1. These values make much more chemical sense.

Abegg's Rule

The range of oxidation states for a specific element is sometimes alluded to in introductory chemistry. As examples, values for sulfur range from −2 to +6, while for chlorine the range is from −1 to +7. It was Abegg who, in 1904, noticed that the sum of the extreme oxidation states of an element often equaled eight. The popularization of this observation did not happen until 1916, in a long-overlooked contribution to chemical bonding by Lewis [32]. The rule can be states as follows:

> **Abegg's Law** states that, for a main group element, the total difference between the maximum negative and positive oxidation states of an element is frequently eight and is in no case more than eight.

As an example, sulfur has the oxidation state limits of −2 and +6. Thus, applying Abegg's law: [{+6} − {−2}] = +8.

Electron Gain and Loss

Having devoted the first part of this chapter to the variously defined concept of electronegativity, the second part will be on the very specifically defined topics of ionization energy and electron affinity. Values of which are mostly known to considerable precision.

First a comment upon a statement that appears in many introductory chemistry texts: "Ionic compounds form because metals want to give up valence electrons and nonmetals want to gain valence electrons." The statement is a convenient fiction for students starting out in chemistry, but nothing could be farther from the truth! This false explanation can be demolished by simply considering the ionization energy (IE_1) and electron affinity (EA_1) of the sodium atom:

$$Na(g) \rightarrow Na^+(g) + e^- ....IE_1 = +496 \text{ kJ·mol}^{-1}$$
$$Na(g) + e^- \rightarrow Na^-(g)..........EA_1 = -53 \text{ MJ·mol}^{-1}$$

As can be seen from the values, sodium actually "wants" to gain an electron not lose one! It is only the fact that the nonmetal counter-atom has a higher electron affinity that "forces" sodium to lose its valence electron. That is, ionic bonding is not benign, but atomic "nature red in tooth and claw," in other words, a competition for the valence electrons [33]. The two related phenomena are discussed in the following.

Ionization Energy

One pattern explicable in terms of electron configuration is that of ionization energy. Usually we are interested in the *1st ionization energy*. As the orbital occupancy may change between the neutral atom and the ionized ion, a correct definition is as follows [34]:

> The experimental **1st ionization energy** is equal to the difference between the total electronic energy of the atom X and the total electronic energy of the ion X^+, both in their ground states. That is, $X(g) \rightarrow X^+(g) + e^-$

Periodic Trends in Ionization Energy

Unlike the molecule-dependent values of covalent radii, ionization energies can be measured with great precision. Figure 2.4 shows the IE_1 for the 1st, 2nd, and 3rd Period elements. As can be seen, the pattern is repetitious, the Group 1 elements at the low point and the Group 18 elements at the peaks. Most of the variations can be explained

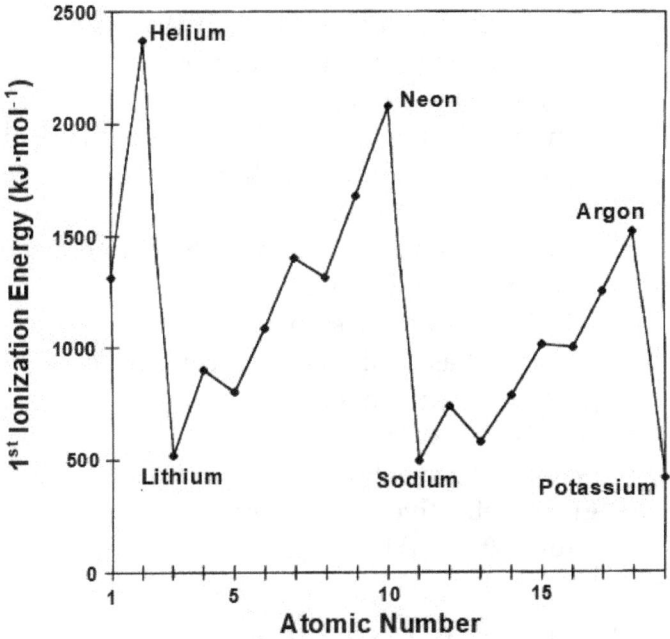

Figure 2.4 1st ionization energy for the first 19 elements (adapted from Ref. [35]).

in terms of screening/shielding from the nucleus of the outermost electron by the inner electrons [35].

Instead of discussing IE_1 of each element shown, one cycle will be chosen for examination: that of the 2nd Period elements. The patterns can be explained as follows:

- Lithium has a small IE_1 as the 2s electron is largely shielded from the nuclear attraction by the $1s^2$:
  $[He]2s^1 \rightarrow [He]$
- Beryllium has a larger IE_1 primarily as a result of the greater effective nuclear charge:
  $[He]2s^2 \rightarrow [He]2s^1$
- Boron has a lower IE_1 as, even though there is an increase in nuclear charge, the 2p electron is partially shielded by the $2s^2$ electrons:
  $[He]2s^22p^1 \rightarrow [He]2s^2$
- Carbon has a higher IE_1 primarily as a result of the greater effective nuclear charge:
  $[He]2s^22p^2 \rightarrow [He]2s^22p^1$
- Nitrogen has a higher IE_1 primarily as a result of the greater effective nuclear charge:
  $[He]2s^22p^3 \rightarrow [He]2s^22p^2$
- Oxygen has a lower IE_1 which will be discussed separately in the following:
  $[He]2s^22p^4 \rightarrow [He]2s^22p^3$
- Fluorine has a higher IE_1 primarily as a result of the greater effective nuclear charge:
  $[He]2s^22p^5 \rightarrow [He]2s^22p^4$
- Neon has a much higher IE_1 primarily as a result of the greater effective nuclear charge:
  $[He]2s^22p^6 \rightarrow [He]2s^22p^5$

The Half-Filled Shell Myth

Ingrained in the vocabulary of chemistry is the term "the stability of the half-filled shell." However, it is not the "stability" of the p^3 configuration, but the reduced "stability" of the subsequent electrons, which accounts for the break in near-linearity of the plot. Cann has compared some of the explanations for the discontinuity and concluded the following one to be the best [35]:

> Because of the Pauli Exclusion Principle, electrons with parallel (unpaired) spins tend to avoid each other, thus decreasing the electrostatic repulsion between them. This will be the situation when filling the first half of the shell. When electrons are forced to doubly occupy orbitals in the second half, their spins are constrained to be paired (antiparallel). Because they are no longer obliged to avoid each other, the [inter-electron] electrostatic repulsion increases.

In Figure 2.5, the IE_1 are shown for the 2p and 3p block elements. Continuing the line of the p^1 to p^3 configurations, a line parallel to the actual p^4 to p^6 values is obtained. The difference between the two represents the coulombic repulsion between pairs of electrons within the same orbital. For the 2p series, this amounts to about 430 kJ·mol^{-1}, while for the 3p series it is 250 kJ·mol^{-1}. Cann attributed the difference

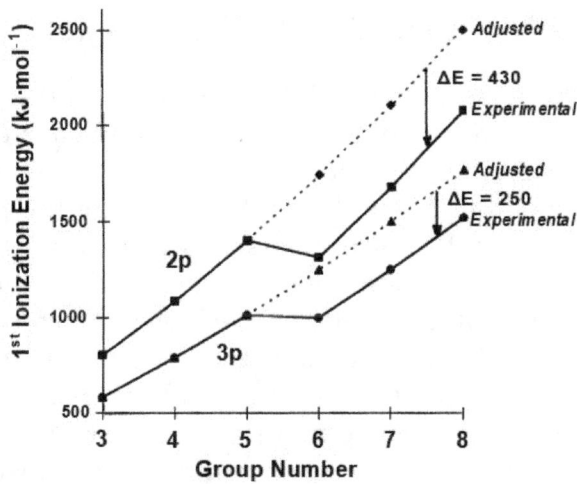

Figure 2.5 First ionization energy for the p-block elements of the 2nd and 3rd Periods (adapted from Ref. [35]).

between the two series to the more diffuse 3p orbitals compared with the 2p orbitals. Thus, any paired 3p electrons are sharing a larger volume of space and therefore have less mutual repulsive forces.

To review, there is nothing exceptional about the "half-filled shell." It is instead the interelectron repulsive forces between the electron pairs beyond the p^3 configuration, which result in a lower ionization energy. To reinforce the point, as Rich and Suter added [36]:

> *Likewise, when one compares the energy to remove an electron from the half-filled p subshell with that needed for a p^2 structure, nothing special is found.*

Rich and Suter then referred to the claimed stability of the "filled shell." They continued [36]:

> *Similarly, the large energy difference between electrons in $3s^1$ and $2p^6$ configurations is readily explained by the difference in principal quantum number; this again indicates no more "extra" stability of a filled p shell than it does for a p^5 or any other structure in which the electron being removed is at the lower principal number.*

3d-Series Metal Ionization Energies

The 1st and 2nd ionization energies of the 3d-series metals, corresponding to the removal of each of the $3s^2$ electrons, show a steady increase without any major deviations [37]. More interesting are the 3rd ionization energies, IE_3, of the 3d-series metals. With these subsequent ionization energies, it is the d electrons that are being "plucked off" one by one. As can be seen from Figure 2.6, it is an almost identically shaped plot to that for the 2p and 3p electrons in Figure 2.5, except that the greater coulombic repulsion between any electron pairs commences with the d^6 configuration (instead of p^4), as expected.

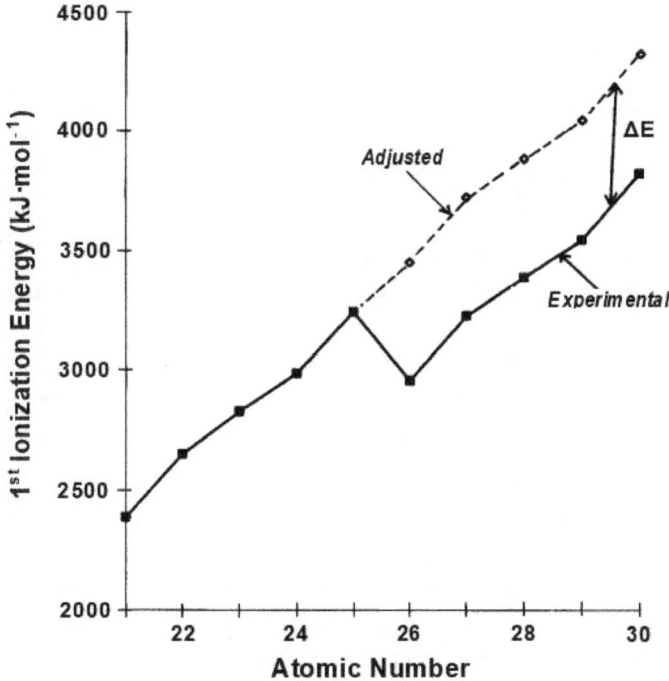

Figure 2.6 3rd ionization energy (IE$_3$) for the 3d-block elements (adapted from Ref. [35]).

Group Trends in Ionization Energy

Proceeding down a group, the 1st ionization energy generally decreases. This is especially systematic for the noble gases.

$He(g) \rightarrow He^+(g) + e^-$ ....IE$_1$ = +2.37 MJ·mol^{-1}
$Ne(g) \rightarrow Ne^+(g) + e^-$ ....IE$_1$ = +2.08 MJ·mol^{-1}
$Ar(g) \rightarrow Ar^+(g) + e^-$ ....IE$_1$ = +1.52 MJ·mol^{-1}
$Kr(g) \rightarrow Kr^+(g) + e^-$ ....IE$_1$ = +1.35 MJ·mol^{-1}
$Xe(g) \rightarrow Xe^+(g) + e^-$ ....IE$_1$ = +1.17 MJ·mol^{-1}

Though the number of protons in the nucleus has increased, so has the number of shielding electron shells. In addition, as the sequential orbitals are filled, the electrons in the outermost shell occupy a larger volume of space and thus have lower interelectrons repulsion factors.

Successive Ionization Energies

There are also patterns in successive ionizations of an element [38]. One of the simplest examples is lithium:

$$Li(g) \rightarrow Li^+(g) + e^- \dots\dots\dots\dots IE_1 = +0.52 \text{ MJ·mol}^{-1}$$
$$Li^+(g) \rightarrow Li^{2+}(g) + e^- \dots\dots\dots\dots IE_2 = +7.30 \text{ MJ·mol}^{-1}$$
$$Li^{2+}(g) \rightarrow Li^{3+}(g) + e^- \dots\dots\dots\dots IE_3 = +11.82 \text{ MJ·mol}^{-1}$$

Lithium has the electron configuration $1s^2 2s^1$. Thus, the first electron to be removed is strongly shielded by the two 1s electrons. Then, to remove each of the 1s electrons requires very much greater energy. The lesser value for removing the second electron compared to the third can be accounted for by two factors: First, there are always electron–electron repulsions when two electrons occupy the same orbital; second, even within the same orbital, one electron does partially shield the other electron.

Electron Affinity

Much space is usually given to ionization energy and little to electron affinity (rarely, but more correctly, called electron attachment energy). Yet as mentioned earlier, atoms usually "want" to gain electrons and certainly not lose them! The following definition is parallel to that given for ionization energy.

*The experimental **1st electron affinity** is equal to the difference between the total electronic energy of the atom X and the total electronic energy of the ion X⁻, both in their ground states. That is, $X(g) + e^- \rightarrow X^-(g)$*

Sign Convention for Electron Affinity

For clarity, it is important to commence with a mention of the confusion over the sign convention for electron affinity. A proponent of the traditional sign convention (no longer in common use) was Wheeler, who contended that [39]:

> With this convention, the electron affinity is positive for elements such as fluorine, for which energy is released when an electron is added to make an ion, while the widely quoted values for the alkaline earth metals and noble gases are negative.

This convention, however, is the opposite of that used for ionization energy. To remove the ambiguity, Brooks *et al.* proposed that the term "electron affinity" should be eliminated and, instead, the reverse process should be regarded as the 0th ionization energy [40]:

$$F^-(g) \rightarrow F(g) + e^- \dots\dots\dots\dots IP_0 = +328 \text{ kJ·mol}^{-1}$$

This format, which never gained wide acceptance, would correspond with the sign convention used here for electron affinity:

$$F(g) + e^- \rightarrow F^-(g) \dots\dots\dots\dots EA_1 = -328 \text{ kJ·mol}^{-1}$$

Period Patterns in Electron Affinity

If anything, the patterns for electron affinity are more interesting than those of ionization energy [41, 42]. The graph in Figure 2.7 shows the first electron attachment energies for the 1st, 2nd, and 3rd Periods.

As with ionization energy, there are the two factors involved: interelectron repulsion and exchange energy. There is still an effective nuclear charge on the periphery of each atom, which increases as the number of protons

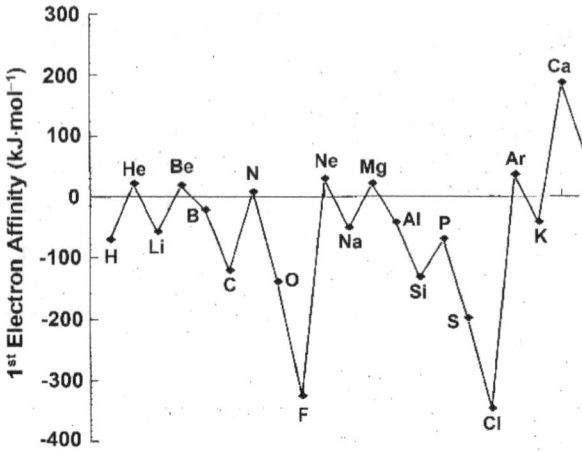

Figure 2.7 Electron affinity (EA_1) hydrogen to calcium.

increases. In the 2nd Period, for example, the greatest EA_1 is that of fluorine. There are three exceptions to the negative EA_1: beryllium, nitrogen, and neon.

- Beryllium has a positive EA_1 as an added electron would have to enter a 2p orbital where it would be shielded by the $2s^2$ electrons. In fact, the electron repulsion must exceed the nuclear attraction:
  $[He]2s^2 \rightarrow [He]2s^22p^1$
- Nitrogen has a positive EA_1 as a result of the interelectronic repulsion being greater than the effective nuclear attraction:
  $[He]2s^22p^3 \rightarrow [He]2s^22p^4$
- Neon has a positive EA_1 as an added electron would have to enter a 3s orbital where it would be shielded from the nuclear attraction particularly by the $2s^2$ and $2p^6$ electrons. In fact, the electron repulsion must exceed the nuclear attraction from the nucleus:
  $[He]2s^22p^6 \rightarrow [He]2s^22p^63s^1$

Group Trends in Electron Affinities

Down a group, as the atoms become larger and the nuclear attraction becomes less, so the electron affinities decrease. The trend is illustrated in Figure 2.8.

The 2nd Period elements from boron to fluorine are clearly exceptions to the trends in their respective groups. Their electron attachment energies are significant deviations from the smooth progressions of the other members of their groups. That is, their electron attraction energy is significantly less than expected. For example, that of nitrogen is +7 kJ·mol^{-1} while that for phosphorus is −72 kJ·mol^{-1}; similarly, that of oxygen is −141 kJ·mol^{-1} while that for sulfur is −200 kJ·mol^{-1}. An accepted explanation is that the atoms are so small that the interelectron repulsion factor is exceptionally

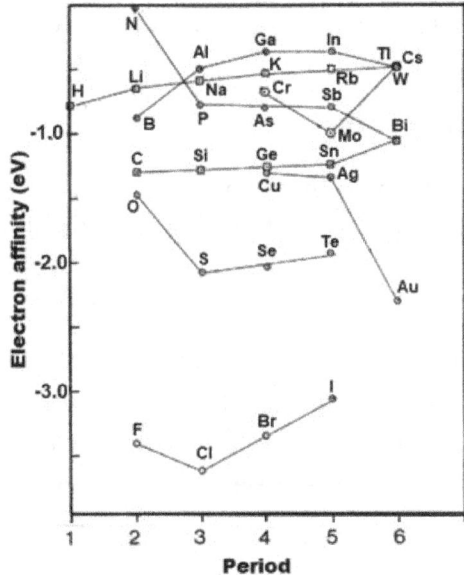

Figure 2.8 A plot of 1st electron affinities by period (adapted from Ref. [41]).

large and, as a result, the attraction for an additional electron is significantly reduced. The anomalous electron affinity of gold will be discussed later in the chapter.

Multiple Electron Affinities

Just as there are multiple ionization energies, so there are the corresponding multiple electron affinities. However, whereas the atomic ionization energies are always positive, as discussed earlier, the 1st electron affinity is often negative. Nevertheless, the subsequent electron affinities are all positive as a result of the increasing electron–electron repulsions. This can be illustrated by the electron affinities of the nitrogen atom:

$$N(g) + e^- \rightarrow N^-(g) \dotfill EA_1 = -7 \text{ kJ·mol}^{-1}$$
$$N^-(g) + e^- \rightarrow N^{2-}(g) \dotfill EA_2 = +673 \text{ kJ·mol}^{-1}$$
$$N^{2-}(g) + e^- \rightarrow N^{3-}(g) \dotfill EA_3 = +1070 \text{ kJ·mol}^{-1}$$

Alkalide Ions

As the formation of the Na^- ion is energetically favored, then compounds containing that ion should be feasible.

$$Na(g) + e^- \rightarrow Na^-(g) \quad EA_1 = -53 \text{ kJ·mol}^{-1}$$

It was in 1974 that Dye *et al.* synthesized the first known compound containing the sodide ion [43]. The team realized that, in the solid phase, there was little energy needed for the formation of the sodium cation–anion pair:

$$2\,Na(s) \rightarrow Na^+(s) + Na^-(s)$$

The key, then, was to find a way of keeping the two ions separated. To do this, Dye *et al.* caged the sodium ion in a bicyclic diaminoether, commonly known as 2,2,2-crypt. The synthesis was successful and gold-colored crystals of $[Na(C_{18}H_{36}N_2O_6)]^+ \cdot Na^-$ were produced. From the crystal structure, the radius of the sodide ion was calculated to be 217 pm, close to that of the iodide ion, and the sodide compound has a structure similar to that of the analogous iodide: $[Na(C_{18}H_{36}N_2O_6)]^+ \cdot I^-$. The preparation of anions of the other alkali metals followed [44]. Then in 1987, Concepcion and Dye synthesized a simpler compound of the sodide ion: $[Li(diaminoethane)_2]^+ \cdot Na^-$ [45].

Since then, simple stable compounds of both the sodide ion and the potasside ion have been synthesized [46]. Of note, the tradition of using the Latin-derived name for the anion was not followed as these anions should have been named "natride" and "kalide," respectively. No explanation was stated, though perhaps it was to avoid confusion of "natride" with "nitride."

A particularly intriguing compound is the so-called "inverse sodium hydride." Sodium hydride itself, Na^+H^-, is a well-known reducing agent as a result of the "naked" hydride ion [47]. By "caging" the hydrogen ion, it has been possible to synthesize $[H^+]_{cage}Na^-$ [48].

The Auride Ion

Looking at the plot of electron affinities (Figure 2.8), gold stands out as an obvious candidate for anion formation.

$$Au(g) + e^- \rightarrow Au^-(g) \quad EA_1 = -223 \text{ kJ} \cdot \text{mol}^{-1}$$

In fact, the first evidence for the formation of an auride came in 1937 by the equimolar mixing of cesium and gold [49]. This transparent yellow compound was shown in

1959 not to be an alloy, but to be Cs^+Au^-, with a sodium chloride crystal structure. Since then, several other auride compounds have been synthesized [50], including tetramethylammonium auride, $[N(CH_3)_4]^+ \cdot Au^-$. The compound is isostructural to the corresponding bromide, which further illustrates the similarities between the auride and halide ions [51].

The Platinide Ion

At -205 kJ·mol^{-1}, EA_1 for platinum is close to that of gold. Thus, it should come as no surprise that there is an increasing chemistry of the platinide ion, Pt^{2-}, including cesium platinide, Cs_2Pt [52].

Relativistic Effects on Atomic Properties

As an explanation for the significantly negative electron affinity, and other anomalous behavior, relativistic effects must be invoked [53]. These effects are rarely discussed in general chemistry [54], yet they are vital to the comprehension of many facets of atomic behavior [55]. Two of the contexts in which relativistic effects are discussed are the color of gold [56, 57] and the liquid phase of mercury at room temperature [58]. In this section, the focus will be on the relativistic explanation for the formation of auride and platinide ions and then in later chapters on some other relevant relativistic phenomena.

Though the electrons in all atoms experience some degree of relativistic effects, they only become important for the heavier elements. There are two significant factors that can be ascribed to relativistic effects [59]

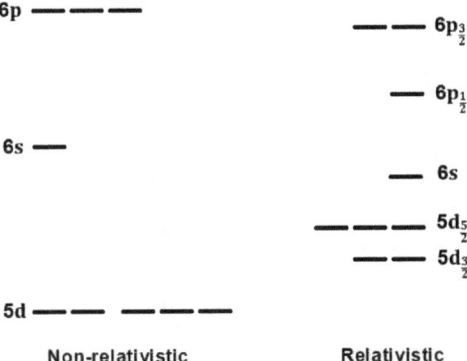

Figure 2.9 Nonrelativistic and relativistic energy levels for the 5d, 6s, and 6p orbitals (adapted from Ref. [59]).

(Figure 2.9 shows both factors for the 5d, 6s, and 6p energy levels):

- **Changing in relative energy levels of atomic orbitals**
  s orbitals decrease substantially in energy and p orbitals decrease to a lesser extent when relativistic effects are taken into consideration. This results in increased shielding of the nucleus, causing d orbitals and f orbitals to increase in energy.

- **Splitting of energy levels having $l > 0$ into two sublevels as a result of spin–orbit coupling**
  p levels split into $p_{1/2}$ and $p_{3/2}$ while the d levels split into $d_{3/2}$ and $d_{5/2}$ levels.

Platinum and Gold Electron Affinities

It is relativistic effects that can explain the high EA_1 for platinum and gold. The additional electron enters the 6s orbital:

$$\text{Platinum: } [\text{Xe}]6s^0 4f^{14} 5d^{10} \rightarrow [\text{Xe}]6s^1 4f^{14} 5d^{10}$$
$$\text{Gold: } [\text{Xe}]6s^1 4f^{14} 5d^{10} \rightarrow [\text{Xe}]6s^2 4f^{14} 5d^{10}$$

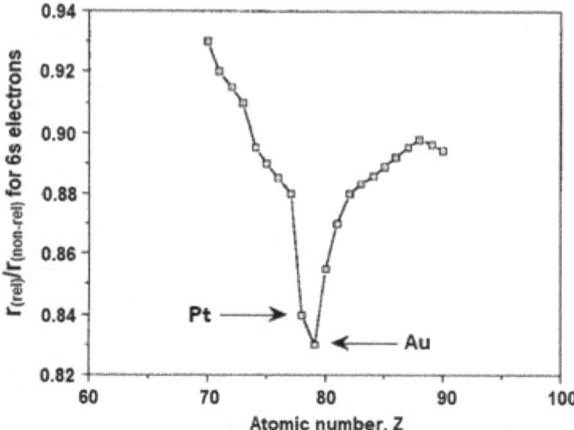

Figure 2.10 Plot of ratio of relativistic to nonrelativistic atom radii for the 6s orbital (adapted from Ref. [60]).

As can be seen from Figure 2.10, the relativistic decrease in relative radius for an added 6s electron reaches a minimum at gold, with the value for platinum being not substantially different [60]. That is, there will be a greater effective nuclear charge on any additional 6s electron for platinum and gold than would be expected without taking relativistic effects into account.

Commentary

In this chapter, a mere selection of atomic periodic properties have been chosen for discussion. In this way, the Reader is not overwhelmed by endless tables and graphs of data. Those who wish to indulge should look elsewhere. This book is designed to make the many concepts of elemental relationships become alive and stimulating, not boring and soporific. The chapter has ended with an introduction to relativistic effects. This oft-overlooked aspect will not be simply a passing reference, but a topic that will be revisited in different contexts in later chapters.

References

1. E. R. Scerri, "What Is an Element? What Is the Periodic Table? And What Does Quantum Mechanics Contribute to the Question?" *Found. Chem.* **14**, 69–81 (2012).
2. R. J. Myers, "What Are Elements and Compounds?" *J. Chem. Educ.* **89**, 832–833 (2012).
3. E. Ghibaudi, A. Regis, and E. Roletto, "What Do Chemists Mean When They Talk About Elements?" *J. Chem. Educ.* **90**, 1626–1631 (2013).
4. W. B. Jensen, "Logic, History, and the Chemical Textbook: II. Can We Unmuddle the Chemistry Textbook?" *J. Chem. Educ.* **75**(7), 817–828 (1998).
5. R. L. Johnson, "The Development of Metallic Behaviour in Clusters," *Phil. Trans. R. Soc. London A* **356**, 211–230 (1998).
6. P. G. Nelson, "Definition of 'Element'," *Chem. Educ. Res. Pract.* **7**(4), 288–289 (2006).
7. W. B. Jensen, "Electronegativity from Avogadro to Pauling: Part I. Origins of the Electronegativity Concept," *J. Chem. Educ.* **73**(1), 11–20 (1996).
8. W. B. Jensen, "Electronegativity from Avogadro to Pauling: Part II. Late Nineteenth- and Early Twentieth-Century Developments," *J. Chem. Educ.* **80**(3), 279–287 (2003).
9. M. R. Leach, "Concerning Electronegativity as a Basic Elemental Property and Why the Periodic Table Is Usually Represented in Its Medium Form," *Found. Chem.* **15**, 13–29 (2013).
10. E. J. Little and M. M. Jones, "A Complete Table of Electronegativities," *J. Chem. Educ.* **37**(5), 231–233 (1960).
11. A. L. Allred and E. G. Rochow, "A Scale of Electronegativity Based on Electrostatic Forces," *J. Inorg. Nucl. Chem.* **5**, 264–268 (1958).
12. W. B. Jensen, "The Quantification of Electronegativity: Some Precursors," *J. Chem. Educ.* **89**, 94–96 (2012).

13. K. Ruthenberg and J. C. M. González, "Electronegativity and Its Multiple Faces: Persistence and Measurement," *Found. Chem.* **19**, 61–75 (2017).
14. H. L. Accorinti, "Incompatible Models in Chemistry: The Case of Electronegativity," *Found. Chem.* **21**, 71–81 (2019).
15. R. T. Sanderson, "Electronegativities in Inorganic Chemistry III," *J. Chem. Educ.* **31**, 238–245 (1954).
16. T. L. Meek and L. D. Garner, "Electronegativity and the Bond Triangle," *J. Chem. Educ.* **82**(2), 325–333 (2005).
17. W. B. Jensen, "The Historical Development of the van Arkel Bond-Type Diagram," *Bull. Hist. Chem.* **13/14**, 47–59 (1992).
18. W. B. Jensen, "A Quantitative van Arkel Diagram," *J. Chem. Educ.* **72**(5), 395–398 (1995).
19. L. C. Allen *et al.*, "Van Arkel-Ketelaar Triangles," *J. Mol. Struct.* **300**, 647–655 (1993).
20. S. S. Ghule *et al.*, "Synthesis, Physical Properties and Band Structure of Non-Magnetic Y_3AlC," *Phys. B* **498**, 98–103 (2016).
21. S. Ullah *et al.*, "Structural, Electronic and Optical Properties of $AgXY_2(X = Al, Ga, In and Y = S, Se, Te)$," *J. Alloys Compd.* **617**, 575–583 (2014).
22. G. Sproul, "Electronegativity and Bond Type: Predicting Bond Type," *J. Chem. Educ.* **78**(3), 387–390 (2001).
23. C. S. McCaw and M. A. Thompson, "A New Approach to Chemistry Education at Pre-University Level," *Nat. Chem.* **1**, 95–96 (2009).
24. J. Šima, "Oxidation Number: Issues of Its Determination and Range," *Found. Chem.* **11**, 135–143 (2009).
25. W. B. Jensen, "The Origin of the Oxidation State Concept," *J. Chem. Educ.* **84**(9), 1418–1419 (2007).
26. J. G. Calvert, "Glossary of Atmospheric Chemistry Terms (Recommendations 1990)," *Pure Appl. Chem.* **62**(11), 2167–2219 (1990).
27. H-P. Loock, "Expanded Definition of the Oxidation State," *J. Chem. Educ.* **88**(3), 282–283 (2011).
28. W. B. Jensen, "Oxidation States versus Oxidation Numbers," *J. Chem. Educ.* **88**(12), 1599–1600 (2011).
29. J. M. Kauffman, "Simple Method for Determination of Oxidation Numbers in Compounds," *J. Chem. Educ.* **63**(6), 474–475 (1986).

30. A. A. Woolf, "Oxidation Numbers and Their Limitations," *J. Chem. Educ.* **65**(1), 45–46 (1988).
31. K. Pavel, P. McArdle, and J. Takats, "Comprehensive Definition of Oxidation State (IUPAC Recommendations 2016)," *Pure Appl. Chem.* **88**(8), 831–839 (2016).
32. G. N. Lewis, "The Atom and the Molecule," *J. Am. Chem. Soc.* **38**(4), 762–785 (1916).
33. R. Schmid, "The Noble Gas Configuration — Not the Driving Force but the Rule of the Game in Chemistry," *J. Chem. Educ.* **80**(8), 931–937 (2003).
34. K. A. Waldron *et al.*, "Screening Percentages Based on Slater Effective Nuclear Charge as a Versatile Tool for Teaching Periodic Trends," *J. Chem. Educ.* **78**(5), 635–639 (2001).
35. P. Cann, "Ionization Energies, Parallel Spins, and the Stability of Half-Filled Shells," *J. Chem. Educ.* **77**(8), 1056–1061 (2000).
36. R. L. Rich and R. W. Suter, "Periodicity and Some Graphical Insights on the Tendency toward Empty, Half-full, and Full Subshells," *J. Chem. Educ.* **65**(8), 702–704 (1988).
37. P. S. Matsumoto, "Trends in Ionization Energy of Transition-Metal Elements," *J. Chem. Educ.* **82**(11), 1660–1661 (2005).
38. P. F. Lang and B. C. Smith, "Ionization Energies of Atoms and Atomic Ions," *J. Chem. Educ.* **80**(8), 938–946 (2003).
39. J. C. Wheeler, "Electron Affinities of the Alkaline Earth Metals and the Sign Convention for Electron Affinity," *J. Chem. Educ.* **74**(1), 123–125 (1997).
40. D. W. Brooks *et al.*, "Electron Affinity: The Zeroth Ionization Potential," *J. Chem. Educ.* **50**(7), 487–488 (1973).
41. E. C. M. Chen and W. E. Wentworth, "The Experimental Values of Electron Affinities: Their Selection and Periodic Behavior," *J. Chem. Educ.* **52**(8), 486–489 (1975).
42. R. T. Myers, "The Periodicity of Electron Affinity," *J. Chem. Educ.* **67**(4), 307–308 (1990).
43. J. L. Dye, "Alkali Metal Anions: An Unusual Oxidation State," *J. Chem. Educ.* **54**(6), 332–339 (1979).
44. J. L. Dye, "Compounds of Alkali Metal Anions," *Angew. Chem. Int. Ed.* **18**, 587–598 (1979).
45. R. Concepcion and J. L. Dye, "Li$^+$(en)$_2$·Na$^-$: A Simple Crystalline Sodide," *J. Am. Chem. Soc.* **109**, 7203–7204 (1987).

46. J. Kim *et al.*, "Crystalline Salts of Na⁻ and K⁻ (Alkalides) That Are Stable at Room Temperature," *J. Am. Chem. Soc.* **121**(45), 10666–10667 (1999).

47. P. C. Too *et al.*, "Hydride Reduction by a Sodium Hydride–Iodide Composite," *Angew. Chem. Int. Ed.* **55**(11), 3719–3723 (2016).

48. M. Y. Redko *et al.*, "'Inverse Sodium Hydride': A Crystalline Salt That Contains H⁺ and Na⁻," *J. Am. Chem. Soc.* **24**(21), 5928–5929 (2002).

49. W. Biltz *et al.*, "Über Wertigkeit und chemische Kompression von Metallen in Verbindung mit Gold," *Z. anorg. allgem. Chem.* **236**(1), 12–23 (1938).

50. M. Jansen, "The Chemistry of Gold as an Anion," *Chem. Soc. Rev.* **37**, 1826–1835 (2008).

51. P. D. C. Dietzela and M. Jensen, "Synthesis and Crystal Structure Determination of Tetramethylammonium Auride," *Chem. Commun.* 2208–2209 (2001).

52. M. Jansen, "Effects of Relativistic Motion of Electrons on the Chemistry of Gold and Platinum," *Solid State Sci.* **7**(12), 1464–1474 (2005).

53. D. R. McKelvey, "Relativistic Effects on Chemical Properties," *J. Chem. Educ.* **60**(2), 112–116 (1983).

54. M. S. Banna, "Relativistic Effects at the Freshman Level," *J. Chem. Educ.* **62**(3), 197–198 (1985).

55. P. Pyykkö, "Relativistic Effects in Structural Chemistry," *Chem. Rev.* **88**(3), 563–594 (1988).

56. A. H. Guerrero, H. J. Fasoli, and J. L. Costa, "Why Gold and Copper Are Colored but Silver Is Not," *J. Chem. Educ.* **76**(2), 200 (1999).

57. P. Schwerdtfeger, "Relativistic Effects in Properties of Gold," *Heteroat. Chem.* **13**(6), 578–584 (2002).

58. L. J. Norrby, "Why Is Mercury a Liquid? Or, Why Do Relativistic Effects Not Get Into Chemistry Textbooks?" *J. Chem. Educ.* **68**(2), 110–113 (1991).

59. J. S. Thayer, "Relativistic Effects and the Chemistry of the Heaviest Main-Group Elements," *J. Chem. Educ.* **82**, 1721–1729 (2005).

60. M. Bardaji and A. Laguna, "Gold Chemistry: The Aurophilic Attraction," *J. Chem. Educ.* **76**(2), 201–203 (1999).

Chapter 3

First Period Problems

The placement of hydrogen in the Periodic Table is still a cause of dissention. In some versions of the Periodic Table, uniquely among the elements, the hydrogen symbol appears twice. Here, the different proposed locations will be described and contrasted. Though all common versions of the Periodic Table place helium in Group 18, some chemists propose its placement should be in Group 2, based upon electron configuration.

Rouvray has lauded the accomplishments of the Periodic Table [1]:

> *The periodic table is also deeply reassuring in that it accounts for and assigns a specific position to every element; even elements still waiting to be synthesized have their rightful place awaiting them.*

This claim is not quite correct, as Cronyn described in 2003 [2]:

> *Quite remarkable, after 130 years of construction, the place of hydrogen in the periodic table is still the subject of doubt, confusion, and inadequate explanation that appears to be little more than numerology.*

Despite Cronyn's somewhat hyperbolic language, it is indeed true that the position of hydrogen remains a contentious issue. So much so, almost this whole chapter is devoted to the problem.

Hydrogen Location: An Overview

First, it should be noted that the hydrogen location problem is partially an artifact from using the rectangular short form or long form of the Periodic Table. Other designs, such as spiral or circular tables, enable the "H" symbol to be placed at the center, making the issue is moot. One such hydrogen-centered Periodic

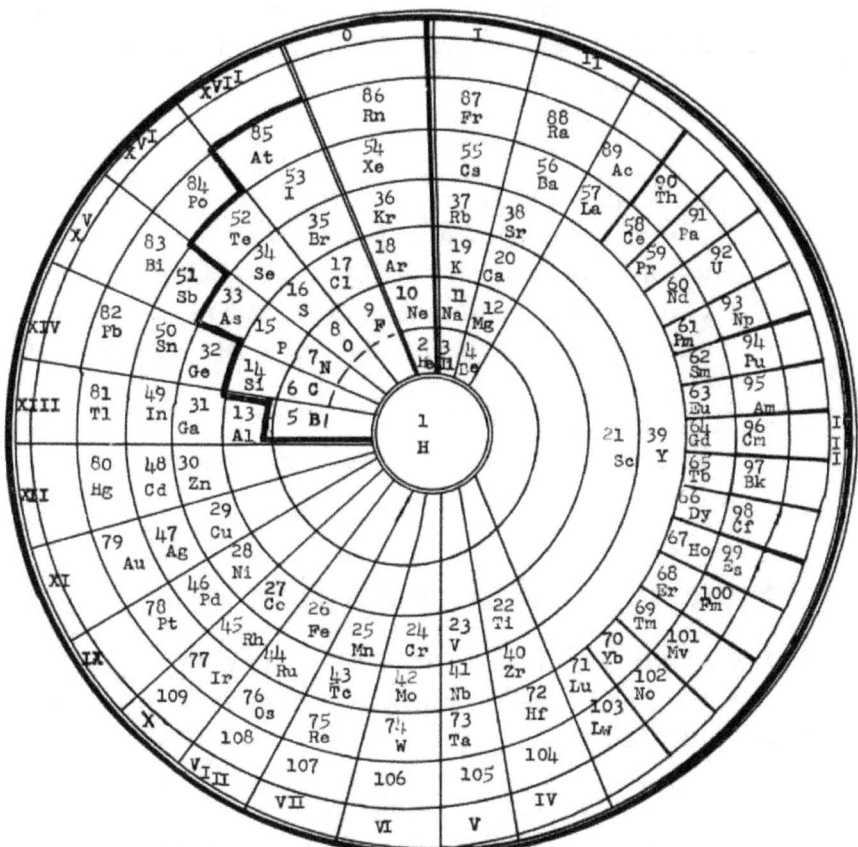

Figure 3.1 Griff's spiral Periodic Table of 1964 with "H" at the center (from Ref. [4]).

Table was proposed by Piutti in 1925 [3] while a more elegant spiral design was drawn by Griff in 1964 [4] (Figure 3.1).

However, as the conventional Periodic Tables are almost ubiquitous, placement in these "standard" tables will be the focus of this chapter. While other elements are largely locked into place by the ordering of the atomic number, the fact that hydrogen and helium are above the next eight-member Period results in this dilemma.

Before discussing the pros and cons of each option, it is useful to the Reader to see the common alternatives listed. Most of these options have been reviewed by Petruševski and Cvetković [5]:

- Hydrogen as a member of Group 1
- Hydrogen as a member of Group 17
- Hydrogen as a member of Group 1 and Group 17
- Hydrogen as a member of Group 13 or 14
- Hydrogen as a member of Group 1 and Group 14 and Group 17
- Hydrogen on its own
- Hydrogen as a member of Group 1 and helium as a member of Group 2

Hydrogen as a Member of Group 1

Most commercial Periodic Tables show hydrogen as a member of Group 1. Cronyn has described the situation elegantly [2]:

> So there is poor hydrogen, denied a chemical family to call its own, thrust like an unwanted orphan into a foster home where its chemistry cannot even be discussed in the same breath with the alkali metals where it now resides.

The strong argument for the placement of hydrogen in Group 1 is that of electron configuration, as championed by Scerri [6]. As the alkali metals have a valence electron configuration of ns^1, then, as the ground state electron configuration of the hydrogen atom is $1s^1$, it follows that hydrogen should be placed at the head of the Group 1 column (Figure 3.2).

There are some supporting chemical arguments. In theory, the hydrogen atom ionizes to give a (positive) hydrogen ion, just as an alkali metal atom ionizes to give an alkali metal cation. Of course, hydrogen does not, in a chemical context, produce a free proton but is, instead, part

H																															He
Li	Be																									B	C	N	O	F	Ne
Na	Mg																									Al	Si	P	S	Cl	Ar
K	Ca																Sc	Ti	V	Cr	Mn	Fe	Co	Ni	Cu	Zn	Ga	Ge	As	Se	Br	Kr
Rb	Sr																Y	Zr	Nb	Mo	Tc	Ru	Rh	Pd	Ag	Cd	In	Sn	Sb	Te	I	Xe
Cs	Ba	La	Ce	Pr	Nd	Pm	Sm	Eu	Gd	Tb	Dy	Ho	Er	Tm	Yb	Lu	Hf	Ta	W	Re	Os	Ir	Pt	Au	Hg	Tl	Pb	Bi	Po	At	Rn
Fr	Ra	Ac	Th	Pa	U	Np	Pu	Am	Cm	Bk	Cf	Es	Fm	Md	No	Lr	Rf	Db	Sg	Bh	Hs	Mt	Ds	Rg							

Figure 3.2 A traditional long form of the Periodic Table with hydrogen in Group 1 (from Ref. [6]).

of a positively charged network: $[H(OH_2)_n]^+$. There is the similarity in that, in aqueous solution, the alkali metal ions, too, are all strongly aquated.

Nevertheless, the alkali metal group form one of the best examples of systematic group trends. To place hydrogen at the top of it implies that hydrogen, itself, is an alkali metal. Yet every comprehensive inorganic book discusses the chemistry of hydrogen in a whole separate unit from that of the alkali metals [7].

Claims of the metallic nature of hydrogen under exceptionally high pressure are used to support the argument that hydrogen is actually an anomalous alkali metal. It was in 1935 that Wigner and Huntingdon were the first to propose on theoretical arguments that under extremes of compression, dihydrogen would undergo a transition to a metallic allotrope [8]. Much high-pressure research has been attempted in order to obtain physical evidence of the existence of metallic hydrogen. In part, such a search relates to the possibility of metallic hydrogen being in the core of the gas giant planets, Jupiter, Saturn, Neptune, and Uranus.

However, the goal of some researchers has been to provide evidence that hydrogen really should be considered as an alkali metal [9]. There has been a claim that solid, metallic hydrogen has indeed been produced [10]. However, does the formation of a metallic allotrope under such extreme conditions really provide evidence that hydrogen belongs to the alkali metal family? Moore has pointed out the fallacy of that argument [11]:

> Boron, oxygen, sulfur, selenium, tellurium, and phosphorus all can be made conductive under pressure, but only in the case of hydrogen is metallization thought to vindicate its predicted properties.

The most intriguing evidence for hydrogen as a member of Group 1 has come from a different route. A compound has been synthesized containing a four-coordinate hydrogen atom that occupies the same site as lithium or sodium [12].

Hydrogen as a Member of Group 17

It can be argued that the $1s^1$ of hydrogen is one electron short of a complete electron shell, just as are the ns^2np^5 electron configurations of the halogens. However, the reasons for hydrogen placement as a member of Group 17 rely more on the chemical arguments. Supporting evidence for placement in Group 17 is that hydrogen, like the halogens, forms a stable diatomic molecule. By contrast with the halogens, dihydrogen is not as reactive as the dihalogens.

Also similarly, hydrogen can form a negative, hydride ion. Sacks argued vociferously for the placement of hydrogen unambiguously in Group 17 [13]:

> A Coulombic model, in which all compounds of hydrogen are treated as hydrides, places hydrogen exclusively as the first member of the halogen family and forms the basis for reconsideration of fundamental concepts in bonding and structures.

But again, similarity in formula masks a completely different chemical behavior. In particular, the hydride ion decomposes violently in the presence of water, unlike the water stability (albeit with hydrolysis in the fluoride case) of the halide ions.

Hydrogen as a Member of Group 1 and Group 17

At this point, the discussion will be paused to summarize in Table 3.1 the pros and cons of hydrogen placement in Group 1 or in Group 17.

Up to now, only the option of hydrogen placement in a single location has been considered. To avoid settling upon one option or the other, some early periodic classifications placed hydrogen above both halogen and alkali metal Groups [14]. An interesting 18-column table of this type was devised by LeRoy in 1931 (Figure 3.3) [15]. Of relevance to discussions in later chapters, LeRoy (and other contemporary chemists) placed boron and aluminum in Group 3A (now Group 3), not Group 3B (now Group 13).

The dominance of hydrogen was repeated more cleanly in a pyramidal Periodic Table with hydrogen at its apex [16]. The dual linkage of hydrogen to lithium and fluorine became very popular. Some commercial Periodic Tables and some of those within textbooks, locate an "H" symbol to head both

Table 3.1   Summary of placement reasons in Group 1 and Group 17

	Group 1 — Alkali Metal	Group 17 — Halogen
Argument for placement	Possesses single s electron	Nonmetal
	Commonly forms H^+ ion	Gas at room temp.
		Forms diatomic molecule
Argument against placement	Gas at room temp.	Rarely forms H^- ion
	Not a metal	Does not possess p electrons
	Not highly oxidizing	Not highly reducing
	Element unreactive with water	Element unreactive with water

			B										A					
IV	V	VI	VII	Transition	I	II	III	IV	V	VI	VII	0	I	II	III			
												H						
													He	Li	Gl	B		
								C	N	O	F	Ne	Na	Mg	Al		
								Si	P	S	Cl	A	K	Ca	Sc		
Ti	V	Cr	Mn	Fe	Co	Ni	Cu	Zn	Ga	Ge	As	Se	Br	Kr	Rb	Sr	Y	
Zr	Cb	Mo	Ma	Ru	Rh	Pd	Ag	Cd	In	Sn	Sb	Te	I	Xe	Cs	Ba	Rare Earths	
Ct	Ta	W	Re	Os	Ir	Pt	Au	Hg	Tl	Pb	Bi	Po	?	Rn	?	Ra	Ac	
Th	Pa	U																

Figure 3.3 The 1931 LeRoy version of the Periodic Table (from Ref. [15]).

Group 1 and Group 17. In fact, many classrooms and lecture halls in schools, colleges, and universities are adorned with this version (Figure 3.4). This "dodging the choice" is actually worse pædogically, with students coming to believe that hydrogen is both an alkali metal and a halogen.

Elemental Dual Identity

Whether the Reader is supportive of the placement of hydrogen in Group 1 and Group 17, it raises an important point that is usually overlooked: that is, can an element be placed in two locations? It is a crucial philosophical point in this book, as in several instances in later chapters, elements seem to fit in more than one location. Though others before him actually adopted a dual location for an element, Rich seems to have been the first to emphasize that dual (or triple) identity/location of an element was a very significant and fundamental conceptual and philosophical break from the idea that each element occupied a single location [17].

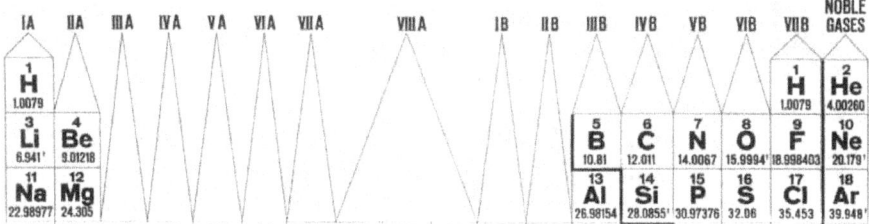

Figure 3.4 The first three Periods of an earlier version of the Fisher Scientific™ wall chart.

Hydrogen as a Member of Group 13 or 14

If the element does not fit well at either extreme of the Periodic Table, where, then? The earliest mention of placing hydrogen in the middle of the Periodic Table was in 1893 by Rang. He devised one of the first 18-member forms of the Periodic Table, numbering the Groups from I to VIII and then I to VII again. He placed the symbol for hydrogen at the head of the second Group III (Figure 3.5). Then in the caption, he noted [18]:

H may not be exactly in its true place, still it cannot be very far from it.

In 1964, Sanderson proposed that hydrogen fitted better in the middle of the Periodic Table, specifically, over carbon. His reasons were that the electronegativity of hydrogen was closer to that of the Group 14 elements and that hydrogen had half-filled outer electron shells. He was careful to suggest that, even though hydrogen should be placed over carbon, it needed to be in a "separate independent position" [19]. Perhaps to make the point unambiguous, his own version of the Periodic Table (Figure 3.6) shows hydrogen bridging over boron and carbon.

Sanderson's choice of placement of hydrogen in Group 14 was supported, and expanded upon, by Cronyn [2]. Cronyn pointed out the similarity in the preference for covalent bond

	I	II	III	IV	V
	"	"	H	"	··
	··	"	"	"	N
	"	"	"	"	P
	Cu	Zn	Ga	Ge	As
	Ag	Cd	In	Sn	Sb
	Au	Hg	Tl	Pb	Bi

Figure 3.5 The center of Rang's 1893 Periodic Table design showing the location for hydrogen (from Ref. [18]).

No.	M 1	M 2	M 2'	M 3	M 4	M 5	M 6	M 7	M 8
1					H 1				He 2
2	Li 3	Be 4		B 5	C 6	N 7	O 8	F 9	Ne 10
3	Na 11	Mg 12		Al 13	Si 14	P 15	S 16	Cl 17	Ar 18

Figure 3.6 Part of Sanderson's 1964 Periodic Table, showing the placement of hydrogen (see Ref. [19]).

formation by both hydrogen and carbon: for example, the H–H bond has a strength of 436 kJ·mol^{-1} while that of the C–H bond is 439 kJ·mol^{-1}. He also commented upon the similarities of the chemistry of hydrogen to that of silicon, cementing the link of hydrogen with Group 14. In his own Periodic Table design, Cronyn reinforced his argument by displaying trends in ionization energy and electron affinity (in eV), showing that the values for hydrogen fitted perfectly in the sequence (Figure 3.7). Electronegativity values were also inserted in Cronyn's Periodic Table, but the value for hydrogen better fitted a pattern for the Group 13 elements.

However, as with the assignment of hydrogen to Group 1 or 17 (or both), pædagogic confusion is caused by a student perception that hydrogen is indeed a formal member of Group 14.

Hydrogen as a Member of Group 1 and Group 14 and Group 17

In all of Laing's Periodic Table proposals, he believed that the Periodic Table was a means of visually displaying chemical

			2.20 **H** 13.60 0.75				— **He** 24.59 —
0.98 **Li** 5.39 0.62	1.57 **Be** 9.32 —	2.04 **B** 8.30 0.28	2.55 **C** 11.26 1.26	3.04 **N** 14.53 —	3.44 **O** 13.62 1.46	3.98 **F** 17.42 3.40	— **Ne** 21.56 —
0.93 **Na** 5.14 0.55	1.31 **Mg** 7.65 —	1.61 **Al** 5.99 0.44	1.90 **Si** 8.15 1.39	2.19 **P** 10.49 0.75	2.58 **S** 10.36 2.08	3.16 **Cl** 12.97 3.61	— **Ar** 15.76 —

Figure 3.7 Part of the Periodic Table by Cronyn showing the values of electronegativity, upper left; ionization energy, lower left; electron affinity, lower right (see Ref. [2]).

linkages and that two or more locations of a single element were educationally beneficial. Rich and Laing suggested that the solution to showing the similarities of hydrogen to each of Group 1, Group 14, and Group 17, was to show hydrogen as a member of each of the three groups (Figure 3.8) [20].

Hydrogen on Its Own

Why should we continue to try to fit hydrogen into a table that is simply a human construct? If hydrogen does not "fit in" perhaps it is because it indeed does not fit in and is best regarded as a unique element. The proposal by Kaesz and Atkins utilized the empty space above the transition metals to place the hydrogen "box" [21]. In doing so, Kaesz and Atkins rejected the two-location model of heading Group 1 and Group 17, stating it had to have a single location. Believing that the chemistry of hydrogen was totally unique, they placed hydrogen central but clearly level with the other 1st Period element, helium (Figure 3.9).

This idea prompted a rapid reply from Scerri who questioned this whole direction of involving observable chemical properties as a factor in the classification of the chemical elements [22]:

A very widely held belief, among chemists and others alike, is that the periodic system consists primarily of a classification of the elements as

1		1				1	2
H		H				H	He

3	4	5	6	7	8	9	10
Li	Be	B	C	N	O	F	Ne

Figure 3.8 The Rich and Laing proposal showing hydrogen as a member of Group 1, Group 14, and Group 17 (see Ref. [20]).

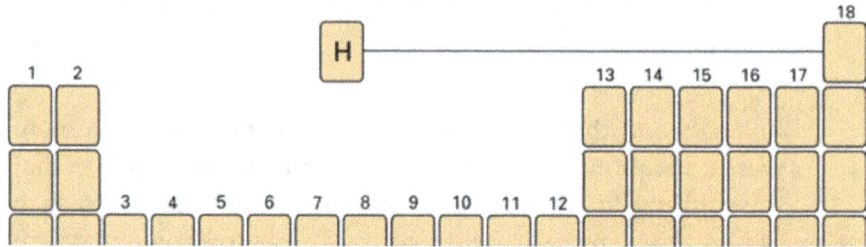

Figure 3.9 The placement of hydrogen according to Kaesz and Atkins (see Ref. [21]).

> simple substances that can be isolated and whose properties can be examined experimentally. However, there is a long-standing tradition of also regarding the elements as unobservable bearers of properties, sometimes elements as basic substances.

To summarize Scerri's critique of Kaesz and Atkins, Scerri believed that the Periodic Table represented the order of atomic structure, not any bulk chemical behavior.

... And Then There Is Helium

Though most of the discussions have centered upon the location of hydrogen, the location of helium has also been contentious.

Hydrogen as a Member of Group 1 and Helium as a Member of Group 2

Up to this point, we have only considered the possible locations of hydrogen in isolation. The arguments were largely, but not entirely, on chemical grounds. If the Periodic Table arrangement is defined by the electron configuration, then logic demands that hydrogen, $1s^1$, and helium, $1s^2$, are placed as the top members of Groups 1 and 2, respectively. One of the Periodic Table designs by Janet in 1928 [23] followed this logic. Then in 1934, White authored a modern-style Periodic Table (reprinted by Laing [24]) displaying electron configurations to reinforce the reason for the design (Figure 3.10).

The first thorough modern discussion of this possibility was given by Katz [25]. In the article, Katz first described moving "He" to above "Be," then shifting Group 1 and Group 2 to the right-hand side of the Periodic Table to generate the left-step (or right-justified) Periodic Table shown in Figure 3.11.

As Katz and, later, Scerri [26] have commented, the left-step table is more elegant than the conventional table. Also the orbitals are now in sequence as f>d>p>s instead of s>f>d>p. The complicating factor arises from electron configuration considerations.

Figure 3.10 The first three Periods of White's 1934 spectroscopic-based Periodic Table (reprinted in Ref. [24]).

																H	He
																Li	Be
									B	C	N	O	F	Ne	Na	Mg
									Al	Si	P	S	Cl	Ar	K	Ca

Sc Ti V Cr Mn Fe Co Ni Cu Zn Ga Ge As Se Br Kr Rb Sr
Y Zr Nb Mo Tc Ru Rh Pd Ag Cd In Sn Sb Te I Xe Cs Ba
La Ce Pr Nd Pm Sm Eu Gd Tb Dy Ho Er Tm Yb Lu Hf Ta W Re Os Ir Pt Au Hg Tl Pb Bi Po At Rn Fr Ra
Ac Th Pa U Np Pu Am Cm Bk Cf Es Fm Md No Lr Rf Db Sg Bh Hs Mt Ds Rg

Figure 3.11 The left-step Periodic Table (from Ref. [6]).

One may protest that helium is not a reactive metal. Bent was not swayed by such an argument. He believed that electron configuration — particularly using the left-step Periodic Table — was the essentiality: the Periodic Table is about atoms, nothing else [27]:

> The answer given here to the Helium Question To Be or Not to Be? is, on both chemical and physical grounds, a resounding Yes! The most noble of the noble gases is not a Noble Gas. Helium's natural position in Periodic Tables is in the s-block above beryllium …

Thus the debate becomes one of chemical properties versus spectroscopic energy levels. Novarro has cited a quote by Scerri that sums up the situation [28]:

> Chemists may place He in the noble gas column, physicists however would rather place it above Be.

In the continuing debate, Ramíriez-Solís and Novarro used quantum-mechanical grounds to argue for helium's place to be above neon [29]. Taking the contrary view, Grochala [30] noted that nothing is seen to be wrong in placing nonmetal hydrogen at the top of the Group 1 metals, so what objection can there be to placing helium also above a metal (beryllium)?

Hydrogen in Group 17 and Helium in Group 18 (Again)

More recently, Scerri has recanted his chemical heresy of placing helium in Group 2 [6]. By combining Period 1 and

																H	He	Li	Be
														B	C	N	O	F	Ne	Na	Mg
														Al	Si	P	S	Cl	Ar	K	Ca
			Sc	Ti	V	Cr	Mn	Fe	Co	Ni	Cu	Zn	Ga	Ge	As	Se	Br	Kr	Rb	Sr
			Y	Zr	Nb	Mo	Tc	Ru	Rh	Pd	Ag	Cd	In	Sn	Sb	Te	I	Xe	Cs	Ba
La	Ce	Pr	Nd	Pm	Sm	Eu	Gd	Tb	Dy	Ho	Er	Tm	Yb	Lu	Hf	Ta	W	Re	Os	Ir	Pt	Au	Hg	Tl	Pb	Bi	Po	At	Rn	Fr	Ra
Ac	Th	Pa	U	Np	Pu	Am	Cm	Bk	Cf	Es	Fm	Md	No	Lr	Rf	Db	Sg	Bh	Hs	Mt	Ds	Rg

Figure 3.12 The revised Scerri left-step Periodic Table (from Ref. [6]).

Period 2 in a single line, helium is safely returned to its chemical "home" atop the noble gases. Hydrogen, meanwhile, is given a home with the halogens, probably a more welcoming location for an element that exists in nature as a diatomic gas, not as a solid metal (Figure 3.12).

Commentary

By 2010, the Editor of the *Journal of Chemical Education* felt that the hydrogen location debate had run its course. In an Editor's Note, Pienta stated [31]:

> ... the Journal is implementing a new policy concerning submissions about the periodic table; those that cover new ground will be considered, but continuing arguments on longstanding issues will not be accepted for review.

As the purpose of this book is to look predominantly at the chemical aspects of the Periodic Table, the convention of placing helium as a member of Group 18 will be followed. For hydrogen, it is the Author's prerogative to choose the location of hydrogen: that will be the Rang, and later Kaesz and Atkins, "lonely" central position.

References

1. D. H. Rouvray, "The Surprising Periodic Table: Ten Remarkable Facts," *Chem. Intell.* **3**, 39–47 (1996).

2. M. W. Cronyn, "The Proper Place for Hydrogen in the Periodic Table," *J. Chem. Educ.* **80**, 947–951 (2003).

3. J. W. van Spronsen, *The Periodic System of Chemical Elements: A History of the First Hundred Years*, Elsevier, Amsterdam, 170 (1969).

4. H. K. Griff, "A Clockwise Spiral System of the Chemical Elements," *J. Chem. Educ.* **41**, 191 (1964).

5. V. M. Petruševski and J. Cvetković, "On the 'True Position' of Hydrogen in the Periodic Table," *Contributions, Sec. Nat. Math. Biotech. Sci. MASA* **38**(1), 83–90 (2017).

6. E. Scerri, "The Role of Triads in the Evolution of the Periodic Table: Past and Present," *J. Chem. Educ.* **85**, 585–589 (2008).

7. M. C. Sneed and R. C. Brasted, *Comprehensive Inorganic Chemistry, Volume 6: The Alkali Metals; Hydrogen and Its Isotopes*, van Nostrand, New York (1965).

8. E. Wigner and H. B. Huntington, "On the Possibility of a Metallic Modification of Hydrogen," *J. Chem. Phys.* **3**, 764 (1935).

9. A. L. Ruoff *et al.*, "Solid Hydrogen at 342 GPA: No Evidence for an Alkali Metal," *Nature* **392**, 46–49 (1998).

10. R. P. Dias and I. F. Silvera, "Observation of the Wigner-Huntington Transition to Metallic Hydrogen," *Science* **355**, 715–718 (2017).

11. J. W. Moore, "Turning the (Periodic) Tables," *J. Chem. Educ.* **80**, 847 (2003).

12. D. M. Cousins, M. G. Davidson, and D. García-Vivó, "Unprecedented Participation of a Four-Coordinate Hydrogen Atom in the Cubane Core of Lithium and Sodium Phenolates," *Chem. Commun.* **49**, 11809–11811 (2013).

13. L. J. Sacks, "Concerning the Position of Hydrogen in the Periodic Table," *Found. Chem.* **8**, 31–35 (2006).

14. G. W. Sears, "A New Form of the Periodic Table as a Practical Means of Correlating the Facts of Chemistry," *J. Chem. Educ.* **1**(8), 173–177 (1924).

15. R. H. LeRoy, "A Modified Periodic Classification of the Elements Adapted to the Teaching of Elementary Chemistry," *J. Chem. Educ.* **8**(10), 2052–2056 (1931).

16. H. A. Wagner and H. S. Booth, "A New Periodic Table," *J. Chem. Educ.* **22**(3), 128–129 (1945).

17. R. L. Rich, "Are Some Elements More Equal than Others?" *J. Chem. Educ.* **82**(12), 1761–1763 (2005).

18. P. J. F. Rang, "The Periodic Arrangement of the Elements," *Chem. News* 178 (14 April 1893).

19. R. T. Sanderson, "A Rational Periodic Table," *J. Chem. Educ.* **41**, 187–189 (1964).

20. R. L. Rich and M. Laing, "Can the Periodic Table Be Improved?" *Educ. Química* **22**, 162–165 (2011).

21. H. Kaesz and P. Atkins, "A Central Position for Hydrogen in the Periodic Table," *Chem. Int.* **25**(6), 1–2 (2003).

22. E. Scerri, "The Placement of Hydrogen in the Periodic Table," *Chem. Int.* **25**(7), 1–3 (2003).

23. P. Stewart, "Charles Janet: Unrecognized Genius of the Periodic System," *Found. Chem.* **12**, 5–15 (2010).

24. M. Laing, "Where to Put Hydrogen in a Periodic Table?" *Found. Chem.* **9**, 127–137 (2007).

25. G. Katz, "The Periodic Table: An Eight Period Table for the 21st Century," *Chem. Educ.* **6**, 324–332 (2001).

26. E. R. Scerri, "Presenting the Left-Step Periodic Table," *Educ. Chem.* **42**, 135–136 (2005).

27. H. Bent, *New Ideas in Chemistry from Fresh Energy for the Periodic Law*, AuthorHouse, Bloomington, Indiana, 117 (2006).

28. O. Novarro, "On the Rightful Place for He within the Periodic Table," *Found. Chem.* **10**, 3–12 (2008).

29. A. Ramíríez-Solís and O. Novarro, "The First Metals in Mendeleev's Table: Further Arguments to Place He above Ne and Not above Be," *Found. Chem.* **16**, 87–91 (2014).

30. W. Grochala, "On the Position of Helium and Neon in the Periodic Table of Elements," *Found. Chem.* **20**, 191–207 (2018).

31. N. J. Pienta, "Editor's Note," *J. Chem. Educ.* **87**, 783 (2010).

Chapter 4

The Group 3 Problem

In Chapter 3, the controversy around the placement of hydrogen (and, to a lesser extent, helium) was described. There is another controversial topic — more so, in fact — of the element choices for the two lower members of Group 3. Here, the issues will be described.

Even more contentious than the placement of hydrogen has been the membership of Group 3. The first two members are self-evident: scandium and yttrium. But which are the subsequent members of the Group? Using the long (32-column) form of the Periodic Table, there is no issue: lutetium and lawrencium fall naturally in place.

However, the 18-column form of the Periodic Table is ubiquitous. The issue becomes which set of 14 elements are pulled out and placed beneath, or whether all 15 elements are placed beneath. Is it the electron configuration or chemistry which should determine which elements occupy these positions? The debate revolved around the third member of the Group: lanthanum or lutetium, as actinium and lawrencium are highly radioactive (hence with short half-lives) and with little established chemistry. In fact, lawrencium was unknown at the time of the early discourse and actinium was assumed to be a transition metal (see Chapter 13) [1].

A History of the Debate

Jensen provided a comprehensive review of the early history of the discourse [2]. In the 1920s and 1930s, lutetium was assigned to Group 3. It was only in the 1940s that the Periodic Table was restructured according to electron configurations. At this time, lanthanum was considered to be the "legal" occupant of the "box" for Period 6, Group 3. Using arguments that will be introduced later, Jensen took the position first proposed by Luder in 1970 [3] that the correct occupants were lutetium and lawrencium, deposing lanthanum and actinium, that were sent to the rows below in the 18-group Periodic Table.

Clark and White reviewed the type of Periodic Table utilized in common U.S. textbooks over the time frame of 1948 to 2008 [4]. They identified, and provided abbreviations for, the three different 18-column Periodic Table formats (displayed in Figure 4.1).

- **14CeTh**. Lanthanum and actinium are placed in Group 3; the 14-member "boxes" commence with cerium (lanthanoids) and thorium (actinoids).
- **14LaAc**. Lutetium and lawrencium are placed in Group 3; the 14-member "boxes" commence with lanthanum (lanthanoids) and actinium (actinoids).
- **15LaAc**. The places in Group 3 are left vacant; the 15-member "boxes" commence with lanthanum and end with lutetium (lanthanoids), then commence with actinium and end with lawrencium (actinoids).

Clark and White showed that from 1948 up until 1984, textbooks were about evenly split between the 15LaAc and 14CeTh formats. It seems to have been Jensen's 1982 article

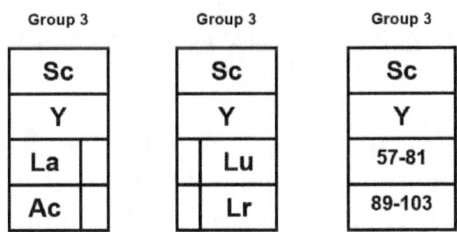

Figure 4.1   The three options for the members of Group 3.

that led to the appearance of, and growth in, the 14LaAc format to the point where it was about an even split between 14CeTh and 14LaAc, the 15LaAc disappearing.

At times, it has seemed less a dispute and more a case of confusion, as Jensen pointed out in 2008 [5]. He noted that the text *Advanced Inorganic Chemistry* by Cotton and Wilkinson used the 14CeTh format on the back flyleaf, but the 15LaAc within the text. Jensen also reported that conversely, Housecroft and Sharpe in the text *Inorganic Chemistry*, used the 15LaAc format on the front flyleaf but the 14CeTh format within the text.

The Dispute Becomes Heated

The years of 2008 and 2009 were busy ones for this debate. In later 2008, Lavelle came to the defense of the former arrangement, arguing that lanthanum and actinium were indeed members of the d-block elements and should remain in the body of the Periodic Table [6].

In a subsequent letter to the *Journal of Chemical Education*, Lavelle referred to the article by Clark and White [7]:

> *In their letter Clark and White wonder why the chemistry education community has not uniformly adopted just one form of the periodic table. Part of the answer is that the majority who are silent on this issue do not want to be attacked by the vocal proponents who insist that lanthanum and actinium must be in the f-block and lutetium and lawrencium must be in the d-block.... I do not wish to enter into conflict but hopefully my article gives voice to those who have been silent. Perhaps our university chemistry textbooks should include brief mention of the difficulties on having one form of the periodic table.*

Clark responded, making the point raised by this Author at the beginning of the chapter. The discourse was based upon the defect of the 18-column form of the Periodic

Table, necessitating 14 elements to be ripped from their rightful place and "dumped" beneath [8]:

... I now favor relegating the flyleaf [18-column] forms to history and shifting the resulting debate to which of the long-form tables is best. There will still be arguments as to which long-form table to use, but these will be healthier arguments than the electron shift agonies of past decades.

The feud continued into 2009. Jensen replied to Lavelle's article [9]:

For obvious reasons I feel compelled to comment on the recent commentary by Lavelle on the placement of La and Ac in the periodic table as I feel that it is not only based on inconsistent reasoning but also contains a serious distortion of the contents of my original article dealing with the subject.

In the same rebuttal, Jensen concluded with [9]:

Finally, with regard to Lavelle's assertion that in his accompanying letter that he speaks for the silent majority who have been cowed into submission by the vocal proponents of the Lu-Lr alternative, I can only say that discussion of this subject is welcome, but for such discussion to be profitable it must be both logically consistent and relevant.

Lavelle's response was published on the following page. He reiterated that, in his view, Jensen's assignments were only accepted by a minority. Lavelle continued [10]:

The point of my discussion on lawrencium was that those who insist on placing lutetium and lawrencium in the d-block, and insist that others do also, are selective in the literature they cite to support the claim.

The personal enmity came through in the closing sentence [10]:

Jensen appears to be unaware of the self-righteous content of some of his articles in this Journal that detract from his otherwise many historically informative publications.

Laing intervened. Jensen, in his 2008 contribution to the debate [5], had cited the Periodic Table from Ephraim's textbook *Periodisches System der Elements*. Laing pointed out that in the 6th English translation, *Fritz Ephraim's Inorganic Chemistry* [11], the lanthanoids were displayed differently. He added [12]:

> In the 25 years that passed between editions, the La-Lu problem was no closer to solution. Another 55 years have now gone by and the debate rages on: plus ça change, plus c'est la même chose.... Arguing about the "right place" for thorium and uranium (or La and Lu or Ac and Lr) in a f- or d-series seems purposeless. Why "should" thorium, [Rn]6d²7s², be best placed in the f-block? ... there is no ideal or perfect periodic table.

Then Scerri joined the debate. In his commentary, he reiterated the point by Clark (and by this Author) [13]:

> It is generally agreed that the conventional or medium-long form table [that is, s-d-p, more often called the short form] continues to survive only because it is more conveniently reproduced in textbooks and wall-charts than the long-form table. The medium-long form table [18-column] relegates as many as 28 elements to a kind of disconnected footnote, and thereby allows one to keep the periodic table relatively slim and having 18 columns. The long-form table, which some textbooks feature, consists of a width of 32 columns. But on the plus side it includes the lanthanoids and actinoids in their rightful place within the main body of the table. More importantly perhaps, it maintains an uninterrupted and increasing sequence of atomic number.

The exchanges ceased, though whether it was because exhaustion had set in, or because the *Journal of Chemical Education* had decided to bar any future acrimonious exchanges (as with the hydrogen placement issue — see Chapter 3), it is not known.

IUPAC Becomes Involved

It was in 2009 that Leigh felt it necessary to clarify the position of the International Union of Pure and Applied Chemistry (IUPAC) on the various periodic table designs in the pages of *Chemistry International* [14]:

> In fact, IUPAC has not approved any specific form of the periodic table, and an IUPAC-approved form does not exist, though even members of

IUPAC themselves have published diagrams titled "IUPAC Periodic Table of the Elements."

Leigh briefly mentioned the Group 3 issue [14]:

As the long form [18-column] places them, the lanthanoids and actinoids sit rather uncomfortably each in a single place, but should lanthanum and actinium be grouped directly with their congeners?

He, too, alluded to the fact that, with what he called the long–long form (here called simply the 32-column form) the Group 3 placement issue did not arise.

Citing Leigh, Scerri argued that IUPAC did need to take a stand on the Group 3 issue [15]:

I propose that IUPAC should in fact take a stance on the membership of particular groups even if this has not been the practice up to this point.

As some 18-member Periodic Tables showed Group 3 as Sc–Y–La–Ac and others Sc–Y–Lu–Lr, Scerri felt the issue needed to be definitively resolved [15]:

This has led to a situation in which chemistry students and professionals alike are often confused as to which version is "more correct" if any.

Scerri pointed out that if the long form of the Periodic Table was written out, then keeping the f- and d-blocks intact, lutetium and lawrencium would naturally occur beneath scandium and yttrium. The Sc–Y–La–Ac alternative could only be structured if, in the long form, the d-block elements of Group 3 (scandium and yttrium) were shifted over next to Group 2, breaking up the transition metal series.

Jensen then rejoined the debate, repeating his 1982 position that lutetium and lawrencium should be assigned to the d-block and lanthanum and actinium as the first members of the f-block. At the end of the abstract, he could not resist the opportunity to continue the feud [16]:

This update is embedded within a detailed analysis of Lavelle's abortive 2008 attempt to discredit this suggestion.

In 2016, an IUPAC task force was instituted to endeavor to resolve definitively which option would be given official IUPAC status [17]. Toward this endeavor, Scerri and Parsons have compiled their own account and recommendations [18].

Some Aspects of the Group 3 Debate

The complexities of the debate would fill a volume on its own. Only some of the key issues will be addressed here. It was unfortunate that the first modern contribution to the discourse, that by Luder, was published in the obscure and long-deceased journal, *Canadian Chemical Education* [3]. In the article, he used two (innovative) 32-column Periodic Tables to show that, by placing lanthanum and actinium in Group 3, Group 3 then became separated from the d-block elements. Instead, it was isolated next to the Group 2 elements. On the other hand, placing lutetium (lawrencium being then unknown) as the third member of Group 3, ensured that these elements became placed next to Group 4. Luder considered the second option the only logical placement. The two options are shown in Figure 4.2.

Physical Properties as Criteria for Location

Jensen's 1982 article [2] raised some interesting points about chemical compatibility. Three of the criteria he chose were atomic radii, the sum of the first two ionization energies, and the melting point. He compared the two options of Sc–Y–La and Sc–Y–Lu with the subsequent d-block sequences of Ti–Zr–Hf, V–Nb–Ta, and Cr–Mo–W (Figure 4.3).

In terms of atomic radii, the early transition metals show a systematic pattern of an increase from 4th Period to 5th Period, then a decrease to the 6th Period. This decrease is a result of the so-called "lanthanoid contraction" (see Chapter 12). The Sc–Y–Lu series follows this pattern, while

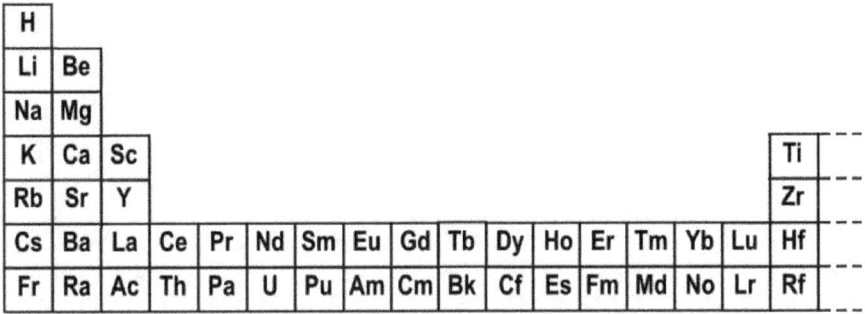

Figure 4.2 The two possible locations for Group 3 according to Luder (Ref. [3]).

the atomic radius of lanthanum is larger than that of yttrium. There is an inverse pattern with the sum of the first two ionization potentials, but then this would be related to atomic radius and is not surprising. For the melting point, the trend is not so precise, though the Sc–Y–Lu series does show an increase (as do the transition metals), while the Sc–Y–La series shows a decrease. From these data, Jensen noted that scandium and yttrium better matched with lutetium. Hence Group 3 was better considered as Sc–Y–Lu–Lr.

In addition to the numerical data, Jensen provided some comparative properties, three of which are shown in Table 4.1. A large range of properties are common to all rare earth elements, but Jensen found a few specific examples

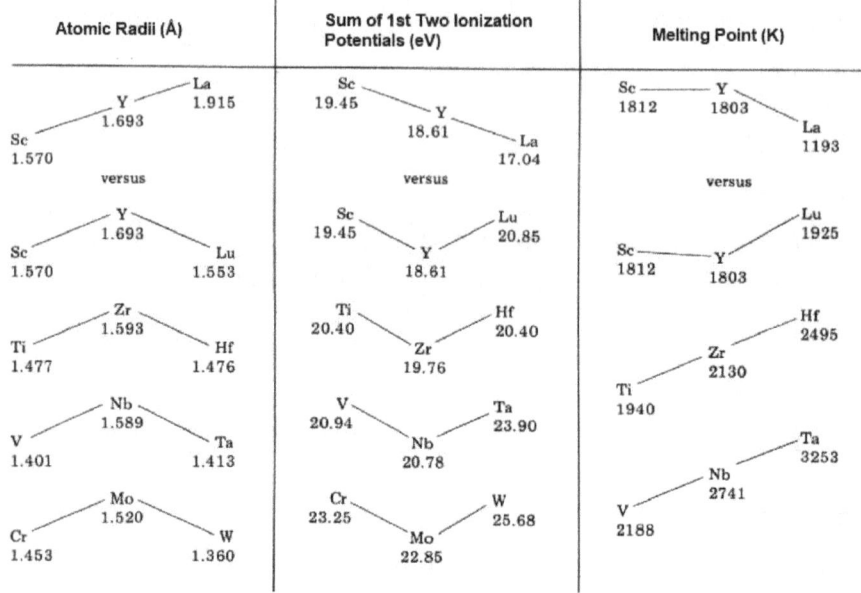

Figure 4.3 Diagrams comparing properties of Group 3 "candidates" to those of early transition metals (adapted from Ref. [2]).

Table 4.1   Some comparative properties of scandium and yttrium with lanthanum and lutetium (adapted from Ref. [1])

Property	Sc	Y	La	Lu
Crystal packing of metal at SATP	Hexagonal close packed	Hexagonal close packed	Double hexagonal close packed	Hexagonal close packed
Highest common oxidation state	+3	+3	+3	+3
Precipitation of sulfate with	Y Group	Y Group	Ce Group	Y Group

where the properties of lanthanum differed from those of scandium, yttrium, and lutetium. Again, he concluded that the better fit was Sc–Y–Lu–Lr.

Endeavoring to identify the key features of the debate, Lavelle's position largely related to electron configurations of the atoms. He argued that as atoms of neither lanthanum nor actinium possessed an electron in an f-orbital, instead possessing an electron in a d-orbital, then both elements needed to be considered as members of the d-block [6]:

> However, placing lanthanum (La) and actinium (Ac) in the f-block is the only case where a pair of elements is placed in a group that results in their being part of a block with no outer electrons in common with that block.

This is shown in Table 4.2, together with the other relevant electron configurations.

The debate discussed in the previous section revolved largely around the relevance of placement of an element upon its ground-state electron configuration. For example, Laing noted that f-block member thorium has instead a ground-state configuration of $[Rn]7s^26d^2$ [12]. The unexpected electron occupancy of a 7p orbital for lawrencium (see Table 4.2) seemed to cause a particularly bitter exchange on the relevance of orbital occupancy for Group 3 membership. In conclusion, though certainly not in closure, to match the

Table 4.2 Electron configurations for the competing elements for membership of Group 3

Sc = $[Ar]4s^23d^1$	
Y = $[Kr]5s^24d^1$	
La = $[Xe]6s^25d^1$	Lu = $[Xe]4f^{14}6s^25d^1$
Ac = $[Rn]7s^26d^1$	Lr = $[Rn]5f^{14}7s^27p^1$

Table 4.3 Comparative electron configurations for the Group 3 (inc. Lu and Lr), Group 4, and Group 5 elements

Group 3	Group 4	Group 5
Sc = $[Ar]4s^23d^1$	Ti = $[Ar]4s^23d^1$	V = $[Ar]4s^23d^1$
Y = $[Kr]5s^24d^1$	Zr = $[Kr]5s^24d^2$	Nb = $[Kr]5s^14d^4$
Lu = $[Xe]4f^{14}6s^25d^1$	Hf = $[Xe]4f^{14}6s^25d^2$	Ta = $[Xe]4f^{14}6s^25d^3$
Lr = $[Rn]5f^{14}7s^27p^1$	Rf = $[Rn]5f^{14}7s^26d^2$	Db = $[Rn]5f^{14}7s^26d^3$

d-block electron configurations (Table 4.3), it would seem most logical to identify the Group 3 elements as Sc–Y–Lu–Lr.

Commentary

In Chapter 5, the term "rare earth metals" will be introduced. By definition, this set of chemical elements includes scandium, yttrium, and all of the elements from lanthanum to lutetium inclusive. Thus, any decision of the membership of Group 3 does not affect the identity of the rare earth metals.

The membership of the f-group elements known as the lanthanoids will be affected, depending upon the definition of Group 3. There is no conflict if all 15 consecutive chemical elements are chosen, which have a common +3 ion charge and which are found in similar ores. Likewise, the actinoids will be a 15-member series. However, if the lower members of Group 3 are excluded from a "double membership" in the appropriate f-block series, then complexities will follow.

In this book, therefore, to avoid placing either of La–Ac or Lu–Lr in both Group 3 and in the f-series in an

18-column Periodic Table, the lower placements in Group 3 will be left empty. It is more important (in this Author's view) to populate the f-series with 15 members and show the continuity of their properties.

References

1. L. S. Foster, "Why Not Modernize the Textbooks Also? I. The Periodic Table," *J. Chem. Educ.* **16**(9), 409–412 (1939).
2. W. B. Jensen, "The Position of Lanthanum [Actinium] and Lutetium [Lawrencium] in the Periodic Table," *J. Chem. Educ.* **59**(8), 634–636 (1982).
3. W. F. Luder, "The Atomic-Structure Chart of the Elements," *Can. Chem. Educ.* **5**(3), 13–16 (1970).
4. R. W. Clark and G. D. White, "The Flyleaf Periodic Table," *J. Chem. Educ.* **85**(4), 497 (2008).
5. W. B. Jensen, "The Periodic Table: Facts or Committees?" *J. Chem. Educ.* **85**(11), 1491–1492 (2008).
6. L. Lavelle, "Lanthanum (La) and Actinium (Ac) Should Remain in the d-Block," *J. Chem. Educ.* **85**(11), 1482–1484 (2008).
7. L. Lavelle, "Response to 'The Flyleaf Periodic Table'," *J. Chem. Educ.* **85**(11), 1491 (2008).
8. R. W. Clark, "Author of 'The Flyleaf Periodic Table' Responds," *J. Chem. Educ.* **85**(11), 1493 (2008).
9. W. B. Jensen, "Misapplying the Periodic Law," *J. Chem. Educ.* **86**(10), 1186 (2009).
10. L. Lavelle, "Response to Misapplying the Periodic Law," *J. Chem. Educ.* **86**(10), 1187 (2009).
11. P. C. L. Thorne and E. R. Roberts (translators), *Fritz Ephraim's Inorganic Chemistry* (6th ed.), Oliver & Boyd, London (1954).
12. M. Laing, "More About the Periodic Table," *J. Chem. Educ.* **86**(10), 1189 (2009).
13. E. Scerri, "Which Elements Belong in Group 3?" *J. Chem. Educ.* **86**(11), 1188 (2009).
14. G. Leigh, "Periodic Tables and IUPAC," *Chem. Int.* **31**(1), 1–2 (2009).

15. E. Scerri, "Mendeleev's Periodic Table Is Finally Completed and What to Do About Group 3?" *Chem. Int.* **34**(4), 1–5 (2012).
16. W. B. Jensen, "The Positions of Lanthanum (Actinium) and Lutetium (Lawrencium) in the Periodic Table: An Update," *Found. Chem.* **17**, 23–31 (2015).
17. E. Scerri (task group chair), "Which Elements Belong in Group 3 of the Periodic Table?" *Chem. Int.* **38**(2), 22–23 (2016).
18. E. Scerri and W. Parsons, "What Elements Belong in Group 3 of the Periodic Table?" In E. Scerri and G. Restrepo (eds.), *Mendeleev to Oganesson: A Multidisciplinary Perspective on the Periodic Table*, Oxford University Press, Oxford, 140–151 (2018).

Chapter 5

Categorizations of the Elements

As will become apparent throughout this book, chemists can be very casual about the use or misuse of chemical terms. This lack of clear definitions even applies to the categorization of the elements themselves. In this chapter, a range of types of categorizations will be introduced and their meanings clarified.

Is polonium a metalloid? What is a weak metal? Which are the noble metals? So many questions, and to many of them, no established definitive answers. In this chapter, definitive proposals will be given as to which elements belong to which categories.

Nonmetals, Metals, and "In-Betweens"

In beginning chemistry, elements are classified as "metals" or "nonmetals." This is a considerable oversimplification, as will be shown in the following. Each element is an individual, but it is convenient for chemists to impose systems of categorization. However, all terminology needs clear definition. Following from the definition, there should be clear reasoning behind assignment of an element to a particular category.

Nonmetals

The nonmetals are a "motley crew." Essentially, they are every element that does not fit into another category. They include all the elements that are in Group 18 and Group 17, plus the top two elements of Group 16 and Group 15, the top element of Group 14, and hydrogen as can be seen in

		Group 14	Group 15	Group 16	Group 17	Group 18
H						He
		C	N	O	F	Ne
			P	S	Cl	Ar
					Br	Kr
					I	Xe
					At	Rn

Figure 5.1 Elements classified as nonmetals.

Figure 5.1. The only contentious element placed here in this category is astatine. Often identified as a metalloid or even a metal, the predominance of evidence is that astatine is a nonmetal [1, 2].

Metalloids

The realization that there was no specific line of demarcation between metals and nonmetals dates back to the late 19th century (even though many commercial Periodic Tables continue to show one) [3]. It was in the 1890s that Newth declared that there were elements with intermediate properties: the metalloids. That was the easy part. The hard part was deciding which elements should be classified as metalloids.

The name "metalloid" was traditionally used for these in-between elements, then the term "semimetal" became preferred. However, the term "semimetal" was subsequently appropriated to be defined in terms of semiconductor materials in general, not simply chemical elements. As a result, "metalloid" has regained its meaning as specifically pertaining to certain chemical elements to the left of the nonmetals in the Periodic Table.

Vernon compiled all the elements identified as metalloids in sources from 1947 until 2012 [4]. The most popular elements cited with their frequency in parentheses were the following nine elements: antimony (88%); arsenic (100%); astatine (40%); boron (86%); germanium (96%); polonium (49%); selenium (23%); silicon (95%); and tellurium (98%). Adapting previously suggested criteria, Vernon devised the following definition for a metalloid:

> A **metalloid** is a chemical element that, in its standard state, has (a) the electronic band structure of a semiconductor or a semimetal, (b) an intermediate first ionization potential (say, 750–1,000 kJ/mol), and (c) an intermediate electronegativity (1.9–2.2, revised Pauling).

According to Vernon, there were six elements only which fitted his criteria for metalloid classification: antimony, arsenic, boron, germanium, silicon, and tellurium.

Hawkes plotted the electrical conductivity (as $\log_{10}$ in $S{\cdot}m^{-1}$) of the proposed metalloids together with a selection of metals and nonmetals along a scale (Figure 5.2). He observed that there were five elements whose electrical conductivity fitted into the gap between metals and nonmetals [5]. These elements were arsenic, boron, germanium, silicon, and tellurium.

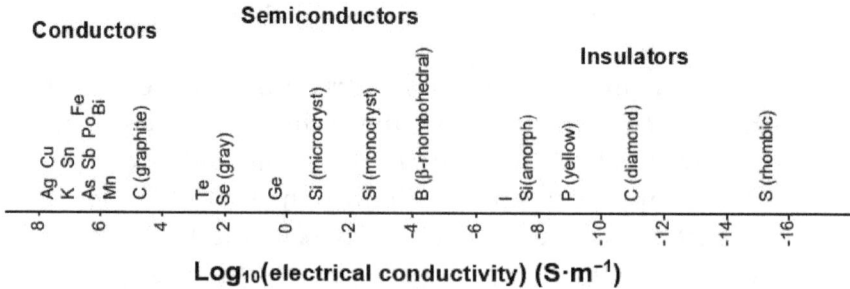

Figure 5.2 A scale of electrical conductivity indicating the placement of some elements (adapted from Ref. [5]).

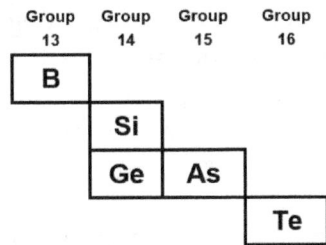

Figure 5.3 The commonly accepted metalloids.

It is this Author's preference to adopt the Hawkes list of five elements as metalloids (Figure 5.3). Antimony, which made Vernon's list but not Hawkes's list, would seem better accommodated in the category in the next subsection (in the following).

Though the categorization earlier is the one that will be adopted here, there is one other candidate for consideration as a metalloid: radon. It has been argued by Stein that radon behaves as an ionic cation in aqueous solution and therefore should be treated as metalloid [6]. An interesting proposal, but one that has not garnered any significant support.

Chemically Weak Metals

To reiterate the point, there are no rigid boundaries in properties across the Periodic Table. Just as we have "invented" an additional category of metalloids for those elements that have properties between metal and nonmetals, so there is now a need for a category between metalloids and "true" metals. A "true" metal has essentially cationic behavior, these metals can also be found as parts of polyatomic anions. A definition is:

> There is a subgroup of the metals, the **chemically weak metals** (or *amphoteric metals* [7]) those closest to the metalloid borderline, that exhibit some chemical behavior more typical of the metalloids, particularly formation of anionic species in basic solution.

The nine elements in this category are aluminum, antimony, beryllium, bismuth, gallium, lead, polonium, tin, and zinc. As an example of anionic species, the pH dependency of zinc ion species is shown in Table 5.1 and compared with that of a "true" metal, magnesium.

Just as zinc ion in very basic solution forms soluble zincates, the other chemically weak metals similarly form aluminates, beryllates, gallates, stannates, plumbates, antimonates, bismuthates, and polonates. The weak metals are shown in Figure 5.4.

The term "chemically weak metals" defines this cluster of elements according to chemical criteria, differentiating them from "normal" metals. To confuse matters, the terms *post-transition metal* and *poor metal* are sometimes used in the literature. However, these categories refer to Periodic Table locations. That is, post-transition metal refers to all

Table 5.1 Variation of species with pH for aluminum and zinc ions

	Very Acidic	Acidic	Basic	Very Basic
Zinc	$Zn^{2+}(aq)$		$Zn(OH)_2(s)$	$[Zn(OH)_4]^{2-}(aq)$
Magnesium	$Mg^{2+}(aq)$		$Mg(OH)_2(s)$	

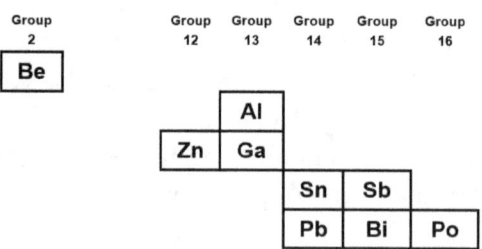

Figure 5.4 The chemically "weak" metals.

the metals of Groups 12 to 16, while poor metals are specifically the metals of the p-block elements (Groups 13 to 16). Also, Habashi devised a category of *less typical metals* that partially overlap with the category here of chemically weak metals [8].

Metals

Though in the teaching of chemistry the focus is usually upon the p-block elements, in fact, about 80% of the naturally occurring elements are metals [9]. All the s-block, d-block, f-block, and the lower left part of the p-block are metals.

The old saying: "I know one when I see one" is often used as a criterion for a metal. As gold prospectors have found to their disappointment "It ain't necessarily so." Among several compounds that have a metallic luster, yellow metallic-looking iron(II) disulfide, FeS_2, mineral name, *pyrite*, well deserves its appellation of "fool's gold." Chemists sometimes refer to a metal by a rather tautological argument as an element containing a metallic bond: that is, bonding throughout the crystal structure involving delocalized electrons [10, 11].

Sometimes metals are defined by a combination of properties, including ductility. Ductility is a measure of a material's ability to undergo significant plastic deformation before rupture. Its opposite, brittleness, is defined as a material that breaks without significant plastic deformation. It is certainly true that some metals, such as gold and lead, are highly ductile, but then other metals, such as beryllium, manganese, uranium, and chromium are very brittle. Similarly, the usually associated term of malleability is true for some elements classed as metals, but not for others.

High three-dimensional electrical conductivity (thus excluding carbon as graphite) is possibly the best superficial indicator of a metallic element [12]. Hawkes has pointed out that under extremes of pressure, the atoms of most elements can be forced into close enough proximity to result in delocalized metallic bonding [13]. Thus in any definition of metals in terms of electrical conductivity, it is important to add "under ambient conditions" or "at SATP." From the best electrical conductor (silver) to the worst (plutonium and manganese) among metals, one is looking at a factor of 10^2 in conductivity difference. Nevertheless, even the worst conducting metals exceed the electrical conductivity of nonmetals and metalloids by a factor of 10^5.

Another reason for stipulating ambient conditions is because the stable allotrope of tin below 13°C, gray α-tin, is nonelectrically conducting. On the other hand, under readily obtainable pressures, iodine becomes electrically conducting. A more specific physical criterion for a metal is the temperature dependence of the electrical conductivity. The conductivity of metals decreases with increasing temperature, whereas that of nonmetals increases.

Supermetals

Metals are commonly accepted as being hard (except mercury), dense, high-melting point, and generally unreactive. At the far left of the conventional Periodic Table, there are metals that are soft, low-melting point, low-density, and highly chemically reactive: the alkali metals. As the alkali metals are so chemically reactive, they deserve their own subcategory: the *supermetals*. If the emphasis is on the high chemical reactivity alone, should the category be broadened to include the three low-density, water-reactive alkaline earth metals? These positive attributes are contradicted by their high-melting points and hardness. The category of supermetals is therefore clearly delineated as being just the alkali metals.

Main Group Appellations

Numerous categorizations have been applied to the chemical elements. The names for Group 1 (*alkali metals*); Group 2 (*alkaline earth metals*); Group 17 (*halogens*); and Group 18 (*noble gases*) have been long accepted. It is curious that for Groups 1 and 2, the term "metals" is appended — using the metal/nonmetal categorization. At the other end, Group 18 is defined as "gases" using the solid/liquid/gas categorization.

For Group 18, at the time of the discovery of the noble gases (when they were more commonly called the "inert gases"), many chemists considered that the so-called "octet rule" precluded compound formation [14]. Now, at least 500 noble gas compounds have been characterized [15]. Adding to the irrelevancy of the "noble gas" name, the latest member, oganesson (element 118), is predicted to be a solid or a liquid at room temperature with a boiling point of between 50°C and 110°C [16]. However, unless very long-lived isotopes are synthesized, it is unlikely that the value can be experimentally confirmed. With ever more noble gas compounds being synthesized, including the exotic Na_2He — actually $(Na^+)_2He(e^-)_2$ [17] — the term "noble" seems ever more inappropriate. As a result, the term *aerogen* is starting to be used. This term first appeared in print in a paper by Noyes, in which he gave credit for the name to Hembold of the University of Oregon [18]. It is now quite widely used in the contemporary literature (see, e.g., Ref. [19]).

Two more group names have become widely adopted. These are *pnictogens* for Group 15 and *chalcogens* for Group 16. The first proposed name for Group 16 was *amphigens* — based on this group's ability to form both acidic and basic compounds — by Berzelius in the early 1800s [20]. The term "chalcogen" was invented by a member of Blitz's research group at the University of Hannover, Germany, about 1932 [21]. Though as a group name, it should apply to all members of Group 16, chalcogen seems to have become commonly used for all Group 16 except for oxygen [22]. Similarly, geochemists refer to chalcophile ("chalcogen-loving") elements. These are the metals and heavier nonmetals that have a low affinity for oxygen and instead react with sulfur, the "heavier" chalcogen, to form insoluble sulfides [23].

The term pnictogen is more recent. The name was devised by van Arkel in the 1950s [24]. International Union of Pure and Applied Chemistry (IUPAC) originally rejected pnictogens as the name for Group 15. In 1970, it was pronounced by IUPAC that if group names were needed, they should be *triels* for Group 13, *tetrels* for Group 14, and *pentels* for Group 15 [25]. Fernelius *et al.* explained [26]:

> There is strong, though not unanimous, sentiment within the IUPAC Commission on the Nomenclature of Inorganic Chemistry to adopt some systematic method for arriving at group names. The following has been suggested.
>
> Family
>
> B, Al, Ga, In, Tl — triels
>
> C, Si, Ge, Sn, Pb — tetrels
>
> N, P, As, Sb, Bi — pentels

Pentels never seemed to be adopted; instead pnictogens (mentioned earlier) became the commonly used term for Group 15 elements. Nevertheless, triels and tetrels have gained in popularity in recent years to describe the Group 13 and Group 14 elements, respectively [27], even in *Chemistry International*, IUPAC's own magazine [28].

d-Block Metal Appellations

There have been numerous names associated with parts of the d-block elements. The issue of which are encompassed by the term "transition metals" (formerly the "transitional metals") will be discussed in Chapter 8. Here are some selected examples that have a specific purpose.

Heavy Transition Metals

Though aficionados of hard rock music may be able to define the term "heavy metal," there is no consensus on the term in the chemical context. In a comprehensive review, Duffus showed that there were four totally different criteria used for the term "heavy metal": high density; atomic mass; atomic number; or toxicity [29]. There was no agreement on cutoff values for any of these criteria. As a result of the ambiguity, Duffus recommended that the term have no official recognition. Hawkes argued that "heavy metal" was a useful descriptor, suggesting that it encompass all metals in Groups 3 to 16 in the 4th Period and beyond [30]. However, this block does seem to include a very large number of disparate metals; and without serving any clear purpose.

By contrast, the similar sounding *heavy transition metals* are well defined [31]. These are the dense large transition metals of the 5th and 6th Periods (Figure 5.5). As will be discussed in Chapter 8, there are many similarities between the lower pair of elements in each transition metal group. The "heavy" term differentiates them from the smaller and significantly lower density transition metals of the 4th Period. Curiously, the companion term "light transition metals" does not seem to have the same degree of popularity.

Group 4	Group 5	Group 6	Group 7	Group 8	Group 9	Group 10	Group 11
Zr	Nb	Mo	Tc	Ru	Rh	Pd	Ag
Hf	Ta	W	Re	Os	Ir	Pt	Au

Figure 5.5 The heavy transition metals.

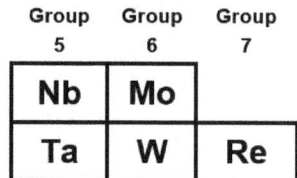

Figure 5.6 The refractory metals.

Refractory Metals

Of significant importance in metallurgy and materials science are the *refractory metals*. The metals fitting this category, shown in Figure 5.6, are niobium and tantalum of Group 5, molybdenum and tungsten of Group 6, and rhenium of Group 7 (if technetium had stable isotopes, it would also fit this category). These metals have melting points above 2000°C and are very hard. In addition, they are chemically inert and have a relatively high density.

Noble Metals

The chemical classification of a *noble metal* is a metal that is highly resistant to oxidation. This category is usually accepted as consisting of the elements: ruthenium and

Group 8	Group 9	Group 10	Group 11
Ru	Rh	Pd	
Os	Ir	Pt	Au

Figure 5.7 The noble metals.

osmium of Group 8; rhodium and iridium of Group 9; palladium and platinum of Group 10; and gold of Group 11 (Figure 5.7).

The synthesis of a compound containing a gold–xenon cation, $[AuXe_4]^{2+}$ [32], ignited an interest in the formation of "noble metal–noble gas" species. However, this raised the issue of which exactly were the noble metals. In a subsequent article on gold–xenon compounds, an analogous mercury(II)–xenon compound was synthesized. This suggested that mercury should also be considered as a noble metal [33]. In another source, it was stated that the noble metals consisted of the platinum metals plus the coinage metals of silver and gold [34]. However, to this Author, the Ru–Os–Rh–Ir–Pd–Pt–Au cluster seems to make the most sense as comprising the "noble metals."

Other Appellations

There are many other terms that have appeared in the context of the chemical elements. In this Author's view, there are three worthy of permanent status; one well known, and two lesser known.

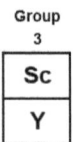

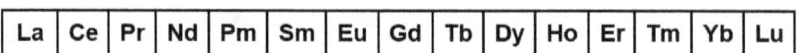

Figure 5.8 The rare earth metals.

Rare Earth Metals

In Chapter 4, the controversy over the membership of Group 3 was described. Fortunately, membership of the *rare earth metals* (also called the *rare earth elements*) is unambiguous. These are the 15 lanthanoid metals plus scandium and yttrium of Group 3 as shown in Figure 5.8. Though they are far from "rare," their ores are found in significant concentrations only in certain specific locations. All of the rare earth metals are characterized by a predominance — though not exclusivity — of the +3 oxidation state, as will be discussed in Chapter 12. Restrepo has shown that in terms of an ordered hypergraph, the rare earth elements do indeed form a complete and unique set of elements [35].

Superheavy Elements

Though the term "heavy metals" is largely meaningless unless defined in context, the use of the term *superheavy elements* has become IUPAC recommended [36]. Commonly abbreviated SHE, the term is given to the postactinoid elements, that is, the elements commencing with rutherfordium, element 104 and continuing to element 126. This limit was chosen presumably in the context of the "magic number" of 126 (see Chapter 1). For any element 127 and onward, the cumbersome term *beyond superheavy elements* was recommended.

Ephemeral Elements

Overlapping — but not identical to — the superheavy elements are the *ephemeral elements* [37]. The concept of ephemerality derives from the Greek word "ephemeros," literally meaning "lasting only one day." The term "ephemeral elements" is not widely used, but it is a useful concept to indicate elements for which there is no known isotope with a half-life of more than a day (hence, following the Greek meaning). In this way, we can roughly identify those elements for which the chemistry is either theoretical, or else based upon the study of individual atoms. At the time of writing, the ephemeral elements are those shown in Figure 5.9.

An isotope of dubnium, element 105, ^{268}Db, has a half-life of 28 hours, thus dubnium is excluded from the ephemeral element list. This list may diminish if longer lived isotopes of any of these elements are synthesized. On the other hand, it will increase if elements beyond $z = 118$ are discovered. Confusingly, the term "ephemeral elements" has been used to describe claims for new elements that were subsequently shown to be false [38].

															At	
No	Lr	Rf		Sg	Bh	Hs	Mt	Ds	Rg	Cn	Nh	Fl	Mc	Lv	Ts	Og

Figure 5.9 The ephemeral elements (at the date of writing).

Commentary

It is quite remarkable that well into the 21st century there are still disagreements on which elements fit into which categories — and the terminology. Hopefully, this chapter has provided some "food for thought."

References

1. S. J. Hawkes, "Polonium and Astatine Are Not Semimetals," *J. Chem. Educ.* **87**(8), 782 (2010).
2. D-C. Sergentu *et al.*, "Advances on the Determination of the Astatine Pourbaix Diagram: Predomination of $AtO(OH)_2^-$ over At^- in Basic Conditions," *Chem. Eur. J.* **22**(9), 2964–2971 (2016).
3. R. H. Goldsmith, "Metalloids," *J. Chem. Educ.* **59**(6), 526–527 (1982).
4. R. E. Vernon, "Which Elements Are Metalloids?" *J. Chem. Educ.* **90**, 1703–1707 (2013).
5. S. J. Hawkes, "Semimetallicity?" *J. Chem. Educ.* **78**(12), 1686–1687 (2001).
6. L. Stein, "New Evidence That Radon Is a Metalloid Element: Ion-Exchange Reactions of Cationic Radon," *J. Chem. Soc. Chem. Commun.* 1631–1632 (1985).
7. C. A. Krause, "The Amphoteric Elements and Their Derivatives: Some of Their Physical and Chemical Properties," *J. Chem. Educ.* **8**(11), 2126–2137 (1931).
8. F. Habashi, "Metals: Typical and Less Typical, Transition and Inner Transition," *Found. Chem.* **12**, 31–39 (2010).
9. P. P. Edwards and M. J. Sienko, "On the Occurrence of Metallic Character in the Periodic Table of the Elements," *J. Chem. Educ.* **60**(9), 691–696 (1983).
10. W. B. Jensen, "The Origin of the Metallic Bond," *J. Chem. Educ.* **86**(3), 278–279 (2009).
11. J. J. Gilman, "In Defense of the Metallic Bond," *J. Chem. Educ.* **76**(10), 1330–1331 (1999).

12. R. T. Myers, "Physical and Chemical Properties and Bonding of Metallic Elements," *J. Chem. Educ.* **56**(11), 712–713 (1979).
13. S. J. Hawkes, "Conductivity," *J. Chem. Educ.* **86**(4), 431 (2009).
14. R. J. Gillespie and E. A. Robinson, "Gilbert N. Lewis and the Chemical Bond: The Electron Pair and the Octet Rule from 1916 to the Present Day," *J. Comput. Chem.* **28**(1), 87–97 (2007).
15. W. Grochala, "Atypical Compounds of Gases, Which Have Been Called 'Noble'," *Chem. Soc. Revs.* **36**(10), 1632–1655 (2007).
16. C. S. Nash, "Atomic and Molecular Properties of Elements 112, 114, and 118," *J. Phys. Chem. A* **109**(15), 3493–3500 (2005).
17. X. Dong *et al.*, "A Stable Compound of Helium and Sodium at High Pressure," *Nat. Chem.* **9**, 440–445 (2017).
18. R. M. Noyes, "Some Predicted Chemistry of Group VIII Elements; the Aerogens," *J. Am. Chem. Soc.* **85**, 22-02-2204 (1963).
19. A. Bauzá and A. Frontera, "Aerogen Bonding Interaction: A New Supramolecular Force?" *Angew. Chem. Int. Ed.* **54**(25), 7340–7343 (2015).
20. W. B. Jensen, "A Note on the Term 'Chalcogen'," *J. Chem. Educ.* **74**(9), 1063–1064 (1997).
21. W. Fischer, "A Second Note on the Term 'Chalcogen'," *J. Chem. Educ.* **78**(10), 1333 (2001).
22. F. A. Devillanova, *Handbook of Chalcogen Chemistry: New Perspectives in Sulfur, Selenium and Tellurium*, Royal Society of Chemistry, London (2006).
23. V. Goldschmidt, "The Principles of Distribution of Chemical Elements in Minerals and Rocks. The Seventh Hugo Müller Lecture, Delivered before the Chemical Society," *J. Chem. Soc.* 655–673 (1937).
24. G. S. Girolami, "Origin of the Terms Pnictogen and Pnictide," *J. Chem. Educ.* **86**(10), 1200–1201.
25. International Union of Pure and Applied Chemistry, *Nomenclature of Inorganic Chemistry, Definitive Rules 1970*, 2nd ed., Butterworths, London, 11, Section 1.2.1 (1970).
26. W. C. Fernelius, K. Loening, and R. M. Adams, "Notes on Nomenclature," *J. Chem. Educ.* **48**(11), 730–731 (1971).
27. L. Brammera, "Halogen Bonding, Chalcogen Bonding, Pnictogen Bonding, Tetrel Bonding: Origins, Current Status and Discussion," *Faraday Discuss.* **203**, 485–507 (2017).
28. P. Metrangolo and G. Resnati, "Categorizing Chalcogen, Pnictogen, and Tetrel Bonds, and Other Interactions Involving

Groups 14–16 Elements," *Chem. Int.* **38**(6), 20–22 (Nov–Dec. 2016).

29. J. H. Duffus, "'Heavy Metals' a Meaningless Term? (IUPAC Technical Report)," *Pure Appl. Chem.* **74**(5), 793–807 (2002).

30. S. J. Hawkes, "What Is a 'Heavy Metal'?" *J. Chem. Educ.* **74**(11), 1374 (1997).

31. S. A. Cotton and F. A. Hart, *The Heavy Transition Elements*, John Wiley, New York (1975).

32. S. Seidel and K. Seppelt, "Xenon as a Complex Ligand: The Tetra Xenono Gold(II) Cation in $AuXe_4^{2+}(Sb_2F_{11}^-)_2$," *Science* **290**(5489), 117–118 (2000).

33. I-C Huang *et al.*, "Gold(I) and Mercury(II) Xenon Complexes," *Angew. Chem. Int. Ed.* **42**, 4392–4395 (2003).

34. M. S. Wickleder and C. Logemann, "2.18 New Chemistry of Noble Metals," in: *Comprehensive Inorganic Chemistry II*, vol. 2, 2nd ed., Elsevier, New York, 491–509 (2013).

35. G. Restrepo, "Challenges for the Periodic Systems of Elements: Chemical, Historical and Mathematical Perspectives," *Chem. Eur. J.* **25**, 15430–15440 (2019).

36. S. Hofmann *et al.*, "On the Discovery of New Elements (IUPAC/IUPAP Provisional Report)," *Pure Appl. Chem.* **90**(11), 1773–1832 (2018).

37. M. W. Browne, "Scientists Meet Analytical Challenge of an Ephemeral Element," *New York Times*, Section C, 3 (8 July 1997).

38. M. Francl, "Ephemeral Elements," *Nat. Chem.* **11**, 2–4 (2019).

Chapter 6

Isoelectronicity

Isoelectronicity is a concept widely used in chemistry — or often misused. It is difficult to think of a more ambiguous term. This chapter commences with a summary of the various ways in which isoelectronic has been used, together with the modifiers that have been applied to delineate its meaning in a specific context. After providing a consistent series of definitions, the richness of isoelectronic relationships, in all their forms, will be explored.

The term *isoelectronic* is ubiquitous in chemistry. But what precisely does a writer mean when they use the term? There is no better way to start the discourse than the immortal words of Humpty Dumpty to Alice, in *Alice Through the Looking Glass*, as shown in Figure 6.1 [1]:

'When I use a word,' Humpty Dumpty said, in rather a scornful tone, 'it means just what I choose it to mean—neither more nor less.'

'The question is,' said Alice, 'whether you can make words mean so many different things.'

Figure 6.1 One of the classic exchanges in *Alice Through the Looking Glass* (from Ref. [1]).

Chemists have shown a strong propensity to use the Humpty Dumpty approach to the word "isoelectronic." Often is the word used, rarely is the word defined. The Reader is left to their own intuition as to which sort of isoelectronicity the author of a specific research paper intended. In this chapter, clarity will be provided by the use of modifiers to remove the need to: ". . . make words mean so many different things."

Historical Definitions of Isoelectronic

The earliest discussion of the isoelectronic concept seems to date from 1919. Langmuir, in his 66-page discourse on the arrangement of electrons in atoms and molecules, commented on patterns among compounds [2]. Though some of the terms he used were archaic, the discussions clearly laid the foundation of the isoelectronic concept. Langmuir employed the symbol, N, to denote the total number of electrons, while E was used to represent the outer valence electrons. He noted the resemblance of carbon monoxide to dinitrogen, and extended the similarity to hydrogen cyanide as they all have N = 14 and E = 10. Langmuir noted many other isoelectronic linkages, such as that of cyanogen, $(CN)_2$, and dichlorine. Of specific note, he saw the predictive ability of the isoelectronic concept [3]:

> For example, since phosphorus and nitrogen atoms contain the same number of electrons in their shells, the simple octet theory represented by Equation 2, indicates that nitrogen compounds corresponding to all known phosphorus compounds could exist and vice versa.

In 1933, Penney and Sutherland [4] made the isoelectronic concept the focus of their work on the shapes of triatomic species, though their definition was somewhat circuituitous: one simply counts the number of electrons in the molecule not in closed electron shells. The authors showed that all triatomics with 16 "molecular or valence" electrons, as they called them, had a linear geometry about the central atom while the 18-electron species were "wide-angled."

Use of the term *isoelectronic principle* seems to have become prevalent in the 1950s and 1960s. Moody [5] used the principle to explain the identical structures of all the following highest oxidation-state fluoride species:

- Group 14: SiF_6^{2-}, GeF_6^{2-}, SnF_6^{2-}
- Group 15: PF_6^-, AsF_6^-, SbF_6^-
- Group 16: SF_6, SeF_6, and TeF_6.

Moody cited Penney and Sutherland for his own definition of isoelectronic:

Simple structures containing the same number of valency electrons are represented by the same bond diagram.

Coulson, in his classic book, *Valence*, described the principle as being [6]:

Two molecules with the same number of valence electrons are isoelectronic: and the principle states that such systems will have similar molecular orbitals.

Brown had a narrower definition [7]:

Two molecular species with the same number of atoms and the same total number of valency electrons are said to be iso-electronic, and the iso-electronic principle states that such molecular species will have similar molecular orbitals and molecular structures.

While companion used isoelectronic in an all-encompassing manner to indicate any cluster of atoms whose total electrons added up to the same value [8]:

The noble-gas atom Ne and the molecules HF, H_2O, NH_3, and CH_4 all have the same number of electrons (are isoelectronic).

Bent had an equally broad definition [9]:

As a general rule, or principle, molecules are isoelectronic with each other when they have the same number of electrons and the same number of heavy-atoms.

Modern Definitions of Isoelectronic

Most first-year university texts use the term "isoelectronic" (without definition) to justify the formation of ions by the main group elements. Specifically, cations are formed that are isoelectronic with the preceding noble gas; anions are formed that are isoelectronic with the following noble gas.

However, there does seem to be some convergence on isoelectronic definitions. For example, Housecroft and Sharpe [10] provided the restrictive definition of:

Two species are isoelectronic if they possess the same total number of electrons.

They continued:

The word isoelectronic is often used in the context of meaning "the same number of valence electrons," although strictly such usage should always be qualified; e.g. HF, HCl, and HBr are isoelectronic with respect to their valence electrons.

While Massey [11] gave a similar but subtly different definition of:

Species that have identical ligands and the same number of electrons on the central atom are said to be isoelectronic and almost invariably they have the same molecular structure.

He, too, looked upon the valence electron counting as a different case:

The principle can often be extended to include species that are not strictly isoelectronic but in which the central atoms have the same number of outer electrons rather than the same total number. The $BeCl_4^{2-}$ and BCl_4^{-}

ions are pseudo-isoelectronic with tetrahedral SiCl₄ and can be expected to have the same structure.

Proposed Definition

Throughout science, the prefix *iso-* means "the same." Thus, strictly speaking, the term "isoelectronic" should simply mean the same number of electrons, period. This all-encompassing meaning needs to be narrowed down if isoelectronic is to have a useful role in identifying chemical patterns and trends. For true (or exact) isoelectronic status, the most logical definition would be the following:

> *Species (atoms, molecules, ions) are **isoelectronic** with each other if they have the same total number of electrons and of valence electrons together with the same number and connectivity of atoms.*

This definition will be used in the following sections of this chapter. One of many more unusual examples of truly isoelectronic species are $[In(NO_3)_4]^-$ and $[Sn(NO_3)_4]$.

Of course, a new term is needed to describe species that have the same number of valence electrons but not necessarily the same total number of electrons. Almost 50 years ago, Gillis had proposed *homoelectronic* [12] while Massey favored the term *pseudo-isoelectronic* [11]. Neither of these terms explicitly identifies their meaning. Elliott and Boldyrev [13] have used the term *valence-isoelectronic*. This would seem the most appropriate term as the Reader is immediately aware of the meaning without need to resort to a chemical dictionary. To provide examples for clarity, OCO and NCO⁻ are isoelectronic, while OCO, OCS, and SCS are valence-isoelectronic. Thus a definition of valence-isoelectronic is:

> *Species (atoms, molecules, ions) are **valence-isoelectronic** with each other if they have the same number of valence electrons together with the same number and connectivity of atoms, but not the same total number of electrons.*

Many examples of valence-isoelectronic species have been identified. Some pairs can be truly startling such as the $[SnBi_3]^{5-}$ ion, obtainable as the potassium compound, which is valence-isoelectronic with the carbonate ion, $[CO_3]^{2-}$ [14].

In addition, as will be discussed in Chapter 9, there are strong similarities between specific compounds in Group (n) and matching compounds in the corresponding Group ($n + 10$). That is, that the compounds specifically differ by a filled d^{10} set (and for the elements lower in the respective groups, there is also a filled f^{14} electron set). It is useful to define this subset of valence-isoelectronic separately. Appropriating Massey's suggestion earlier, *pseudo-isoelectronic* is proposed.

> *Species (atoms, molecules, ions) are **pseudo-isoelectronic** with each other if they have the same number of valence electrons together with the same number and connectivity of atoms, but are differentiated by a d^{10} or $f^{14}d^{10}$ electron set.*

A good example is that of the dioxo cations: CrO_2^{2+}, MoO_2^{2+}, WO_2^{2+}, and UO_2^{2+} [15]. That is, the formula resemblance among the Group 6 ions continues into the pseudo-isoelectronic uranium analogue.

Using the perchlorate ion as an example, Table 6.1 shows an isoelectronic ion (sulfate); a valence-isoelectronic ion (perbromate); and a pseudo-isoelectronic ion (permanganate).

Table 6.1 **Examples of the different subsets of isoelectronicity for the perchlorate ion**

ClO_4^-		
Isoelectronic	Valence-Isoelectronic	Pseudo-Isoelectronic
SO_4^{2-}	BrO_4^-	MnO_4^-

Significance of Isoelectronic Series

There are three reasons why the study of isoelectronic series is important.

- First, it reminds us that, following covalent bond formation, an atom does not "remember" whether, as an element, it was a metal, metalloid, or nonmetal. As such, it can be part of an isoelectronic series across "boundary lines."
- Second, it can be a means of identifying "missing" or additional members of an isoelectronic series and spur the search for synthetic means to prepare them and report their existence.
- Third, the use of isoelectronic substitution can be used to study changes in bonding characteristics.

Following from the first point, the following triad containing a 12-member ring, is an example of an unusual structure for which the nonoxygen atom can be a metal (aluminum), a metalloid (silicon), or a nonmetal (phosphorus). Described by Greenwood and Earnshaw [16], this series is $[Al_6O_{18}]^{18-}$; $[Si_6O_{18}]^{12-}$; and $[P_6O_{18}]^{6-}$ (see Figure 6.2). Sadly, there is no "S_6O_{18}" to complete the set. Sulfur proves to be the exception to the rule, with the analogous sulfur ring compound having half the number of atoms: S_3O_9.

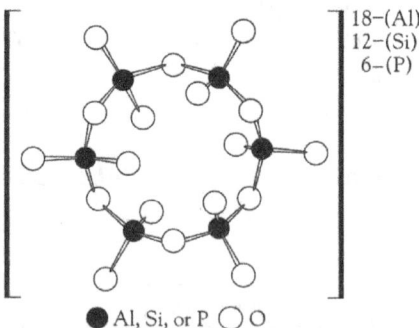

18–(Al)
12–(Si)
6–(P)

● Al, Si, or P ○ O

Figure 6.2 The common $[X_6O_{18}]^{n-}$ isoelectronic ring structure.

fit the formula sequence; however, while the other species are symmetric, these two are asymmetric. This difference can be explained simplistically in terms of the central atom usually being of lower electronegativity. All of these species exhibit delocalized multiple-bond character, including the N_2F^+ ion that has some multiple-bond character in the N–F bond [21]. The "missing" BO_2^- ion does exist, but it is polymeric, not a multiple-bonded monomer. The linear species O_3^{2+} is known in the gas phase. In subsequent tables, transient species produced in gas-phase reactions have been excluded.

Of course, there are many more options if three-element combinations are included. For the 22/16 set in Table 6.4, among those species that can be added are FCN, CNO^-, and even NBC^{4-}.

Five-Atom Isoelectronic Arrays

The isoelectronic array in Table 6.5 shows the stepwise replacement of oxygen atoms by fluorine atoms. In this example, it is the 50/32-electron combinations of 3rd Period elements from Group 14 to Group 17 with oxygen and fluorine. All of the compounds are essentially isostructural, their shape based on the tetrahedron.

Table 6.5 An isoelectronic array of triatomic 3rd Period, Group 14–17 species

# of F/O	Group 14	Group 15	Group 16	Group 17
0/4	SiO_4^{4-}	PO_4^{3-}	SO_4^{2-}	ClO_4^-
1/3	SiO_3F^{3-}	PO_3F^{2-}	SO_3F^-	ClO_3F
2/2	$SiO_2F_2^{2-}$	PO_2F_2	SO_2F_2	$ClO_2F_2^+$
3/1	$SiOF_3^-$	POF_3		
4/0	SiF_4	PF_4^+		

Sequential Isoelectronic Series

It is also possible to construct informative series in which only an individual row is isoelectronic, but successive rows are linked in a simple stepwise manner. For example, in an isoelectronic oxidation-state array, each row contains species having one more electron than the preceding row. Thus, descending the table, the oxidation state of each central atom decreases by one unit.

The array in Table 6.6 shows the 2nd Period series of oxo-species from Group 14 to Group 16. As the oxidation state of the central atom decreases, so the bond angles decrease from 180° to progressively smaller values as one, two, and three nonbonding electrons are added to the central atom. Nitrogen dioxide is one of the few stable radical species, and it is of note that the isoelectronic CO_2^- is stable enough to be found in biological systems [22].

In the matrices and arrays shown earlier, the number of atoms remains the same. Some interesting arrays can be created in which the number of peripheral atoms is decreased stepwise. Table 6.7 shows the second Period 10/8 hydride isoelectronic series for Group 13 to Group 17. The horizontal axis tracks the Group number while the vertical axis represents the decreasing number of hydrogen atoms. Šima has used comparisons of atomic orbital energies to

Table 6.6 Sequential isoelectronic array of triatomic 2nd Period, Group 14–16 species

Electrons	Group 14	Group 15	Group 16	Bond Angle
22/16	CO_2	NO_2^+		180°
23/17	CO_2^-	NO_2	O_3^+	≈135°
24/18		NO_2^-	O_3	≈116°
25/19			O_3^-	114°

Table 6.7 Sequential isoelectronic array of a 2nd Period hydride series (Group 13–17)

# of H	Group 13	Group 14	Group 15	Group 16	Group 17
4	BH_4^-	CH_4	NH_4^+		
3	BH_3^{2-}	CH_3^-	NH_3	H_3O^+	
2			NH_2^-	H_2O	H_2F^+
1				OH^-	HF

Table 6.8 Sequential isoelectronic arrays of chloro-species of 3rd Period, Groups 13–16

# of Cl	Group 13	Group 14	Group 15	Group 16	Geometry
6		$SiCl_6^{2-}$	PCl_6^-		Octahedron
5		$SiCl_5^-$	PCl_5	SCl_5^+	Trig. bipyramid
4	$AlCl_4^-$	$SiCl_4$	PCl_4^+		Tetrahedron

examine why two "missing" members cannot exist, namely H_4O^{2+} and HNe^+ [23].

The following array (Table 6.8) shows the successive isoelectronic rows of 3rd Period main-group chloro-species as chlorine atoms are subtracted. In each column, the element is in its highest oxidation state. Vertically, the geometry changes from octahedral through trigonal bipyramidal to tetrahedral.

A Transition Metal Array

Up to this point, all the discussions have been on arrays involving main-group elements. Arrays can be found, too, for transition metals. Among the transition metals, the

Table 6.9 **Sequential isoelectronic array of fluoro-species of 5th Period, Groups 4–7**

# of F	Group 4	Group 5	Group 6	Group 7
8	HfF_8^{4-}	TaF_8^{3-}	WF_8^{2-}	ReF_8^{-}
7	HfF_7^{3-}	TaF_7^{2-}	WF_7^{-}	ReF_7
6	HfF_6^{2-}	TaF_6^{-}	WF_6	
5	HfF_5^{-}	TaF_5		
4	HfF_4			

"heavy" transition metals show some of the most interesting isoelectronic patterns. The array in Table 6.9 has each 5th Period early transition metal in its highest oxidation state. Descending the table, the number of fluorine atoms decreases until it matches the oxidation state.

Oxidation State as a Variable

In the preceding arrays, each of the central atoms was in their highest oxidation state. It is possible to construct arrays of isoelectronic series in which the variable is not only the number of peripheral atoms, but also the oxidation state of the central atom. This type of array can be illustrated using three successive isoelectronic series of 5th Period fluoro-compounds, stretching across from Group 13 all the way to Group 18 (Table 6.10). These species differ by one charge unit horizontally and two charge units vertically.

There is clearly a "missing" member from the array: the trifluoroxenate(II) ion, XeF_3^{-}. This species has indeed been sought. However, at the time of writing, the only evidence of this ion's existence is as a transient intermediate in gas-phase studies [24].

Table 6.10 Sequential isoelectronic arrays of fluoro-species of 5th Period, Groups 13–18

# of F	Group 13	Group 14	Group 15	Group 16	Group 17	Group 18
6	InF_6^{3-} (+3)	SnF_6^{2-} (+4)	SbF_6^- (+5)	TeF_6 (+6)	IF_6^+ (+7)	
5			SbF_5^{2-} (+3)	TeF_5^- (+4)	IF_5 (+5)	XeF_5^+ (+6)
4					IF_4^- (+3)	XeF_4 (+4)

Arrays of Organometallic Species

A significant proportion of organometallic species obey the 18-(valence)-electron rule [25]. Thus, it is not surprising that there are many possible isoelectronic series in this branch of chemistry. In Table 6.11, each row contains an isoelectronic series of 4th Period transition metal carbonyls with each subsequent row having one carbonyl ligand less [26].

As dinitrogen is, itself, isoelectronic with carbon monoxide, substituting dinitrogen for carbonyl ligands results in another isoelectronic series:

$$Cr(CO)_6, Cr(CO)_5(N_2), Cr(CO)_4(N_2)_2$$

While stepwise substituting nitrosyl for carbonyl requires shifting from metal to metal to maintain isoelectronic status:

$$Ni(CO)_4, Co(CO)_3NO, Fe(CO)_2(NO)_2, Mn(CO)(NO)_2, Cr(NO)_4$$

The isoelectronic principle has also been used to consider, as a replacement for carbon monoxide as a ligand, the potential isoelectronic species of BF and the valence-isoelectronic species of SiO [27]. However, the simple application of isoelectronicity to the trio BF, CO, N_2, does not take into account the polarity and bond order changes along the

Table 6.11 Isoelectronically related organometallic 4th Period carbonyl species (adapted from Ref. [21])

# of CO	Group 4	Group 5	Group 6	Group 7	Group 8	Group 9	Group 10
6	$[Ti(CO)_6]^{2-}$	$[V(CO)_6]^-$	$Cr(CO)_6$	$[Mn(CO)_6]^+$			
5		$[V(CO)_5]^{3-}$	$[Cr(CO)_5]^{2-}$	$[Mn(CO)_5]^-$	$Fe(CO)_5$	$[Co(CO)_5]^+$	
4			$[Cr(CO)_4]^{4-}$	$[Mn(CO)_4]^{3-}$	$[Fe(CO)_4]^{2-}$	$[Co(CO)_4]^-$	$Ni(CO)_4$

series, making simple ligand replacement by BF highly unlikely [28]. Nevertheless, performing a valence-isoelectronic substitution of CO by CS as a ligand has been accomplished [29].

Isoelectronicity: The Future

The isoelectronic principle continues to fascinate. For example, in fullerene research, $C_{59}N^+$ has been synthesized, isoelectronic with C_{60} [30]. A new fruitful area of isoelectronic species is high pressure, high temperature synthesis [31]. One of the early compounds to be manufactured in this category was diboron oxide, B_2O. This compound has a similar structure to the isoelectronic graphite allotrope of carbon [32].

Pyykkö has been extremely active in searching for new and novel isoelectronic series. In the abstract of his review, he wrote [33]:

A combination of ab initio calculations with the isoelectronic principle and chemical intuition is a useful way to predict new species.

Pyykkö was particularly interested in the isoelectronic series of $[PAuP]^{5-}$; $[SAuS]^{3-}$; and $[ClAuCl]^-$. He mused whether the series could be continued to the right. Indeed, valence-isoelectronic $[XeAuXe]^+$ has been identified by mass spectrometry.

Commentary

Clearly, the term "isoelectronic" is a useful one but it is essential that a common definition is agreed. It does seem to make sense to provide a very narrow and unique definition of isoelectronic while valence-isoelectronic can be used for the more general term. As shown earlier, the Reader can see that true (exactly) isoelectronic series "lurk" not only across the nonmetallic elements of each period but even stretch through the metalloid members, into the weak metals. And with the options of valence-isoelectronic, and pseudo-isoelectronic, even more vistas await.

Back in 1952, Coulson ended the section on isoelectronicity with the comment [6]:

> *The isoelectronic principle is not now greatly used except in atomic spectra, and there are, indeed, sometimes difficulties in its application.*

Coulson's gloomy prognosis has proved to be completely wrong. Who knows what other possibilities await to be synthesized?

References

1. L. Carroll, *More Annotated Alice: Alice's Adventures in Wonderland and Through the Looking Glass and What Alice Found There*, with notes by Martin Gardner; Random House, New York, 253 (1990).
2. I. Langmuir, "The Arrangement of Electrons in Atoms and Molecules," *J. Am. Chem. Soc.* **41**, 868–934 (1919).
3. Ref 2, Langmuir, p. 927.
4. W. G. Penney and G. B. B. M. Sutherland, "The Relation between the Form, Force Constants and Vibrational Frequencies of Triatomic Molecules," *Proc. R. Soc. (London)* **A156**, 654–678 (1936).
5. B. Moody, *Comparative Inorganic Chemistry*, 2nd ed., Edward Arnold, London, 51 (1969).
6. C. A. Coulson, *Valence*, Oxford University Press, Oxford, 106 (1952).

7. G. I. Brown, *A New Guide to Modern Valence Theory*, SI ed., Longman, London, 96 (1972).

8. A. L. Companion, *Chemical Bonding*, 2nd ed., McGraw-Hill, New York, 68 (1979).

9. H. A. Bent, "Isoelectronic Systems," *J. Chem. Educ.* **43**(4), 170–186 (1966).

10. C. E. Housecraft and A. G. Sharp, *Inorganic Chemistry*, 2nd ed., Pearson Education, Harlow, UK, 43 (2005).

11. A. G. Massey, *Main Group Chemistry*, 2nd ed., John Wiley, Chichester, 10 (2000).

12. R. G. Gillis, "Isoelectronic Molecules: The Effect of Number of Outer-Shell Electrons on Structure," *J. Chem. Educ.* **35**(2), 66–68 (1958).

13. B. M. Elliott and A. I. Boldyrev, "Ozonic Acid and Its Ionic Salts: Ab Initio Probing of the O_4^{2-} Dianion," *Inorg. Chem.* **43**, 4109–4011 (2004).

14. K. Mayer *et al.*, "[SnBi$_3$]$^{5-}$ — A Carbonate Analogue Comprising Exclusively Metal Atoms," *Angew. Chem. Int. Ed.* **56**(47), 15159–15163 (2017).

15. J. Selbin, "Metal Oxocations," *J. Chem. Educ.* **41**(2), 86–92 (1964).

16. N. N. Greenwood and A. Earnshaw, *Chemistry of the Elements*, 2nd ed., Butterworth-Heinemann, Oxford (1997).

17. R. Lindh *et al.*, "On the Thermodynamic Stability of ArO$_4$," *J. Phys. Chem. A* **103**, 8295–8302 (1999).

18. H. S. Singh, "Oxidations of Organic Compounds with Osmium Tetroxide," in W. J. Mijs and C. R. H. I. de Jonge (eds.), *Organic Syntheses by Oxidation with Metal Compounds*, Springer, Boston, 633–693 (1986).

19. A. J. Ashe, III, "The Group 5 Heterobenzenes," *Acc. Chem. Res.* **11**, 153–157 (1978).

20. M. Larsson *et al.*, "X-Ray Photoelectron, Auger Electron and Ion Fragment Spectra of O_2 and Potential Curves of O_2^{2+}," *J. Phys. B. At. Mol. Opt. Phys.* **23**, 1175–1195 (1990).

21. F. M. Bickelhaupt, R. L. DeKock, and E. J. Baerends, "The Short N–F Bond in N_2F^+ and How Pauli Repulsion Influences Bond Lengths," *J. Am. Chem. Soc.* **124**, 1500–1505 (2002).

22. L. B. LaCagnin *et al.*, "The Carbon Dioxide Anion Radical Adduct in the Perfused Rat Liver," *Mol. Pharmacol.* **33**, 351–357 (1988).

23. J. Šima, "Isoelectronic Series: The Stability of Their Members," *J. Chem. Educ.* **72**, 310–311 (1995).

24. N. Vasdev *et al.*, "NMR Spectroscopic Evidence for the Intermediacy of XeF_3^- in XeF_2/F^- Exchange, Attempted Syntheses and Thermochemistry of XeF_3^- Salts, and Theoretical Studies of the XeF_3^- Anion," *Inorg. Chem.* **49**(19), 8997–9004 (2010).

25. P. R. Mitchell and R. V. Parish, "The Eighteen-Electron Rule," *J. Chem. Educ.* **46**(12), 811–814 (1969).

26. J. E. Huheey, E. A. Keiter, and R. L. Keiter, *Inorganic Chemistry: Principles of Structure and Reactivity*, 4th ed., HarperCollins, New York, 640 (1993).

27. U. Radius *et al.*, "Is CO a Special Ligand in Organometallic Chemistry? Theoretical Investigation of AB, $Fe(CO)_4AB$, and $Fe(AB)_5$ (AB = N_2, CO, BF, SiO)," *Inorg. Chem.* **37**(5), 1080–1090 (1998).

28. R. J. Martinie *et al.*, "Bond Order and Chemical Properties of BF, CO, and N_2," *J. Chem. Educ.* **88**, 1094–1097 (2011).

29. R. B. King *et al.*, "Structural Changes Upon Replacing Carbonyl Groups with Thiocarbonyl Groups in First Row Transition Metal Derivatives: New Insights," *Phys. Chem. Chem. Phys.* **14**, 14743–14755 (2012).

30. K.-C. Kim, F. Hauke, and A. Hirsch, "Synthesis of the $C_{59}N^+$ Carbocation. A Monomeric Azafullerene Isoelectronic to C_{60}," *J. Am. Chem. Soc.* **125**, 4024–4025 (2003).

31. P. F. McMillan, "Chemistry of Materials under Extreme High Pressure-High-Temperature Conditions," *Chem. Commun.* (8), 919–923 (2003).

32. H. T. Hall and L. A. Compton, "Group IV Analogs and the High Pressure, High Temperature Synthesis of B_2O," *Inorg. Chem.* **4**(8), 1213–1216 (1965).

33. P. Pyykkö, "Predicting New, Simple Inorganic Species by Quantum Chemical Calculations: Some Successes," *Phys. Chem. Chem. Phys.* **14**, 14734–14742 (2012).

Chapter 7

Group and Period Patterns among the Main Group Elements

As will be seen in the subsequent chapters, there are a variety of linkages among the elements. In this chapter, the focus will be upon patterns and trends within a Group or a Period. This topic, by necessity, must be limited to a few selected examples.

Whole monographs have been written on patterns and trends down groups and across periods [1–4]. As a result, this chapter on the main group elements will be very selective in the chosen examples.

Main Group Elements

In the 18-Group or 32-Group Periodic Tables, the unfortunate main group elements are fissioned into two widely separated halves: Groups 1 and 2, and Groups 13 to 18. Yet there is no wide gap between beryllium and boron or between magnesium and aluminum. This is an artifact of our obsession with linearity following electron configurations rather than producing a design that has chemical pedagogical value.

Sanderson produced a design to solve this (and other) issues, which was updated by Jensen. Jensen commented [5]:

> *Despite its extraordinary advantages, Sanderson's double-appendix table has seen virtually no use beyond his own writings. It is unclear whether this is due to resistance on the part of authors and publishers, who fear that any departure from the norm will diminish the sale of their textbooks, or to the fact that the use of the periodic table to correlate the facts of descriptive chemistry is so superficial in most textbooks that the very real limitations of the 18-column block table never become apparent.*

A very specific advantage from this Author's perspective is that zinc, cadmium, and mercury are unambiguously

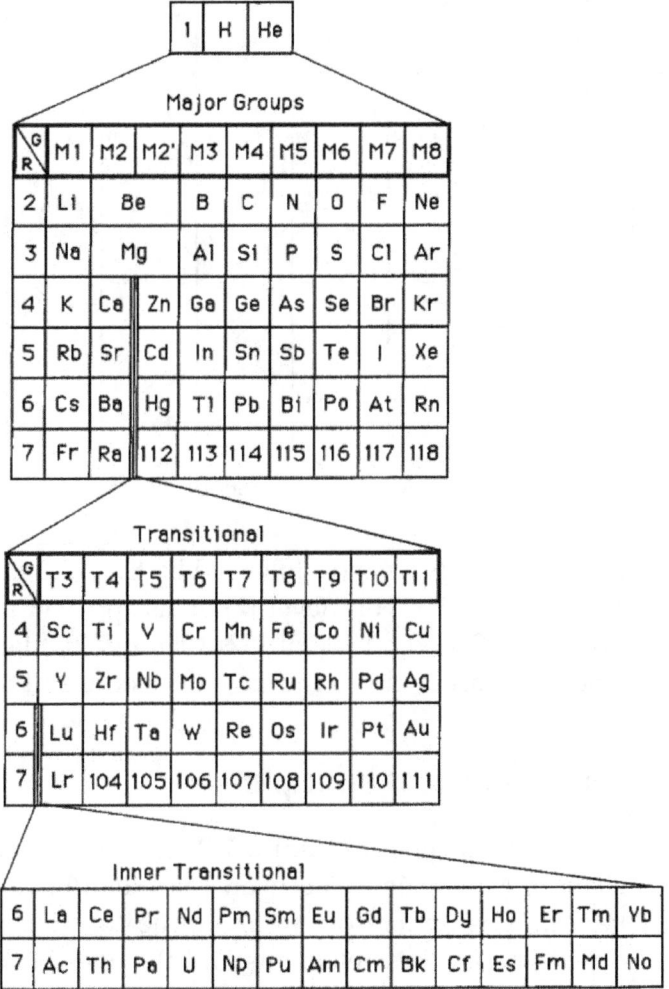

Figure 7.1 Sanderson's double-appendix Periodic Table design (from Ref. [5]).

placed among the main group elements [5]. The Sanderson Periodic Table is shown in Figure 7.1.

Historical Background

It was the concept of repetitive properties that gave rise first to the proposal for triads of elements by Döbereiner, then the Law

Reihen	Gruppe I. R^2O	Gruppe II. RO	Gruppe III. R^2O^3	Gruppe IV. RH^4 RO^2	Gruppe V. RH^3 R^2O^5	Gruppe VI. RH^2 RO^3	Gruppe VII. RH R^2O^7	Gruppe VIII. RO^4
1	H=1							
2	Li=7	Be=9.4	B=11	C=12	N=14	O=16	F=19	
3	Na=23	Mg=24	Al=27.3	Si=28	P=31	S=32	Cl=35.5	
4	K=39	Ca=40	—=44	Ti=48	V=51	Cr=52	Mn=55	Fe=56, Co=59, Ni=59, Cu=63.
5	(Cu=63)	Zn=65	—=68	—=72	As=75	Se=78	Br=80	
6	Rb=85	Sr=87	?Yt=88	Zr=90	Nb=94	Mo=96	—=100	Ru=104, Rh=104, Pd=106, Ag=108.
7	(Ag=108)	Cd=112	In=113	Sn=118	Sb=122	Te=125	J=127	
8	Cs=133	Ba=137	?Di=138	?Ce=140	—	—	—	— — — —
9	(—)		—	—	—	—	—	
10	—	—	?Er=178	?La=180	Ta=182	W=184	—	Os=195, Ir=197, Pt=198, Au=199.
11	(Au=199)	Hg=200	Tl=204	Pb=207	Bi=208	—	—	— — — —
12	—	—	—	Th=231	—	U=240	—	— — — —

Figure 7.2   Mendeléev's Periodic Table of 1871.

of Octaves by Newlands, followed by the periodic patterns of Mendeléev [6] and of Meyer [7]. As is apparent from his version of the Periodic Table of 1871 (Figure 7.2), Mendeléev was guided in part by the formulas of the oxides and the hydrides.

The Uniqueness Principle

As can be seen from Figure 7.2, and of relevance to this chapter, Mendeléev placed an underline beneath the 2nd Period. This underscore was to indicate a degree of difference between these elements and the ones beneath. It was apparent to him that the elements above the line did not wholly resemble the lower members of each Group.

This difference of the topmost member of each Group is known as the *Uniqueness Principle*. Rogers has identified three factors that cause significantly different behavior among most of the 2nd Period elements. Summarizing his statements, it can be stated that [8]:

- They have exceptionally small atomic radii
- They exhibit a maximum of four bonding directions

- The nonmetallic elements have an enhanced ability to form multiple (π) bonds

The theoretical underpinning of the Uniqueness Principle has been described by Kaupp [9]:

> *The similarity of the radial extent of 2s- and 2p-shells is a decisive factor that determines the special role of the 2p elements within the p-block . . . The overall small size of the 2p-elements can be appreciated from any tabulation of atomic, ionic, and covalent radii. . . . This of course does lead to overall coordination number preferences of the 2p-element, as is well known. The smaller radii and the high electronegativities of the 2p-elements are behind their strict obedience of the octet rule, in contrast to the apparent behavior of their heavier homologues . . .*

The Uniqueness Principle can be illustrated by the formulas of the highest oxidation-state simple oxo-anions for the Group 14 and Group 15 elements. The top member has a preference for the delocalized trigonal-planar π-bonding system (Table 7.1), while the lower tetrahedrally coordinated tetra-oxo-anions are prone to polymerization.

Main Group Organometallic Compounds

Until the latter part of the 20th century, it was assumed that formation of organometallic compounds, particularly those with metal–carbon bonds, was an almost exclusive domain of the transition metals. This belief is no longer held. As Power has described [10]:

> *The new compounds that were synthesized highlighted the fundamental differences between their [the heavier main-group elements] electronic properties and those of the lighter elements to a degree which was not previously apparent. This has lead to new structural and bonding insights as well as a gradually increasing realization that the chemistry of the heavier main-group elements ever more resembles that of transition metal complexes than that of their lighter main-group congeners. The similarity is underlined by recent work, which has shown that many of the new compounds react with small molecules such as H_2, NH_3, C_2H_4 or CO under mild conditions and display potential for applications in catalysis.*

Table 7.1   Comparison of the formulas of the simple highest oxidation-state oxo-anions

	Group 14	Group 15
2nd Period	CO_3^{2-}	NO_3^-
3rd Period	SiO_4^{4-}	PO_4^{3-}
4th Period	GeO_4^{4-}	AsO_4^{3-}

The d-Block Contraction

Just as the Uniqueness Principle served to differentiate the chemistry of the 2nd Period elements from those in the subsequent periods, so the *d-block contraction* (sometimes called the *scandide contraction*) results in a greater similarity between some of the members of the 3rd Period and 4th Period [11].

The contraction is not actually a "contraction," instead, the expected increase in effective ionic radii upon descending a p-block element is less when passing from the 3rd Period to the 4th Period than would be expected for a systematic trend. Table 7.2 shows that the increase in radius from Al^{3+} to Ga^{3+} is only 8.5 pm. For comparison, the Group 3 analogue, Sc^{3+}, which like Al^{3+} has a noble gas ion configuration has a radius 21 pm greater. Similarly, the ionic radius of In^{3+} is 10 pm less than that of Y^{3+}.

The accepted explanation for the d-block contraction is that the 3d electrons are poor shielders. Thus the increased effective nuclear charge results in an orbital contraction for gallium and its 3+ ion. The Uniqueness Principle is used to account for the exceptional differences between the 2nd Period and 3rd Period elements of the same Group. Relatedly, the d-block contraction is used to explain exceptional similarities between the 3rd Period and 4th Period elements of the same Group.

Table 7.2   A comparison of the ionic radii of the Group 3 and Group 13 ions

	Group 3	3+ Ion Electron Configuration	M^{3+} (pm)	Group 13	3+ Ion Electron Configuration	M^{3+} (pm)
3rd Period				Aluminum	[Ne]	53.5
4th Period	Scandium	[Ar]	74.5	Gallium	$[Ar]3d^{10}$	62
5th Period	Yttrium	[Kr]	90	Indium	$[Kr]4d^{10}$	80

The 4th Period Anomaly

Following from the d-block contraction is the *4th Period Anomaly*:

> The **4th Period anomaly** *for the p-block elements is where the properties of the Group member of the 4th Period do not fit the trend for the other members of the Group.*

Pyykkö [12] has referred to the phenomenon as *secondary periodicity*, a pattern first reported by Biron in 1915. According to Biron, descending a Group, many physical and chemical properties exhibit an alternating "sawtooth" pattern. Hildebrand rediscovered this phenomenon in the context of the Group 15 elements in 1941 [13]. He noted that the chemistry of nitrogen, arsenic, and bismuth was more focused toward the +3 oxidation state. By contrast, the chemistry of phosphorus and antimony revolved more around the +5 oxidation state.

Dasent wrote an article on compounds that a chemist would expect should exist, but which had not been synthesized at that time. He categorized the probable reasons for nonexistent compounds; category 3 being that of the 4th Period anomaly [14]:

> . . . *those whose instability is related to the reluctance of certain atoms of the first long period to assume their highest oxidation state.*

It was Sanderson who provided a fuller account of the unique features that he found for 4th Period p-block elements, confirming the existence of this anomaly [15]. As exemplars of the 4th Period anomaly, he cited the difficulty in preparing $AsCl_5$ (yet PCl_5 and $SbCl_5$ are well known) [16] and $HBrO_4$ (yet $HClO_4$ and H_5IO_6 are well known) [17].

In explanation, Pyykkö found that there were two different factors. The 4th Period anomaly resulted from the d-block contraction while the lesser 5th Period factor was a result of the combination of relativistic effects (see Chapter 2) and the lanthanoid contraction (see Chapter 12).

Group Trends

In textbook discussions of periodicity, trends in properties within each main group is the most common. An appropriate definition is:

Groups trends are the systematic variation of properties of elements and their compounds descending a specific group. Exceptions to such trends are usually indicative of a change in bonding type.

The large majority of chemical elements are high-melting, unreactive metals. Were the Periodic Table completely filled with these elements, few young people would rush to become chemists! The fascination comes from the s-block and p-block elements where the curious student encounters exotica: highly reactive metals; a yellow solid; a green gas; and so on. This is the diverse world of the main group elements.

It is with the main groups that we have real groups — each with five or six elements for which patterns and trends can truly be traced. Yet each element is unique. Any discussion therefore needs a blend of pattern and individuality.

Here such a blend will be attempted. To do so, the essential features of each element is provided. Why is this necessary? Many/most inorganic texts seem devoid of any sense that chemistry is anything other than theory and calculation. It is now 50 years since Davenport bemoaned the abandonment of the joy of inorganic chemistry [18]:

> ... the typical senior inorganic course leans heavily on theory, particularly bonding theory. Since so many of their teachers are children of the fabled Renaissance of Inorganic Chemistry (surely reports of its implied death were greatly exaggerated?) this is not surprising . . . — is it wise?

Group 1 (Alkali Metals)

Group 1, solely consisting of metals, is one of the few groups to actually show systematic changes descending the series. In this case, for example, chemical reactivity and density increases down the Group.

Sodium, the Second Atypical Alkali Metal

The "abnormality" of lithium is commonly discussed, yet the difference of sodium, also, from the heavier alkali metals is often overlooked [19]. All of potassium, rubidium, and cesium form dioxides(−1), that is, MO_2. Instead, sodium forms a dioxide(−2), Na_2O_2. As another example of the difference, potassium, rubidium, and cesium form triiodide(−en1) compounds, MI_3, whereas lithium and sodium do not.

Zmaczynski [19] has pointed out that sodium compounds with di- and trinegative anions tend to be highly hydrated, such as $Na_2SO_4{\cdot}10H_2O$, $Na_2CO_3{\cdot}10H_2O$, and $Na_2HPO_4{\cdot}12H_2O$. By contrast, the potassium (and rubidium and cesium) compounds are all anhydrous: K_2SO_4, K_2CO_3, and K_2HPO_4.

So significant are the differences of lithium and sodium from the heavier alkali metals, that Smith, in his classic 1917 text, *Introduction to Inorganic Chemistry*, discussed lithium and sodium separately from potassium, rubidium, and cesium [20]. A century later, in 2018, a review article by Restrepo of phenomenological studies included two relevant fragmented Periodic Tables. These Tables, one by Restrepo *et al.* and the other by Leal *et al.*, show lithium and sodium as a separate unit in chemical behavior from the lower three heavier alkali metals [21].

Naked Radii and Hydrated Radii

As chemists, the term "ionic radius" is very clearly defined. Data tables list the values. In the world of biochemistry, the value is larger and fluid. The hydrated ionic radius of an ion is significantly larger than that of the "naked" ion. And it is the inverse order for the alkali metal ion. This results in a free hydrated sodium ion radius of 276 pm compared with 116 pm for the naked ion, while the hydrated radius for potassium is 232 pm compared with 152 for the naked ion. The reason for this can be explained in terms of charge densities. The charge density of the sodium ion is about twice that of the potassium ion. That is, the sodium ion will attract more polar water molecules to it in hydration shells than will potassium. Even though both ions can shed some of the water molecules to pass through passages, in general, the potassium ion will actually pass through many cell wall channels more readily than the sodium ion [22].

Group 2 (Alkaline Earth Metals)

This Group is the first one encountered in which there is only a smooth transition of properties if the first member of the Group is ignored. Thus, from magnesium to barium, chemical reactivity and density increase. Beryllium has a higher density than magnesium, and it exhibits weak metal behavior such as forming beryllates in very basic conditions.

Dolomite: The Mystery Mineral

Containing both calcium and magnesium in precisely equi-molar proportions, *dolomite* has the formula: $CaMg(CO_3)_2$. Massive sedimentary deposits occur on Earth, including those in the Dolomite Alps in Italy. Yet, until recently, when chemists tried to synthesize the compound in the laboratory, all they obtained was a mixture of crystals of magnesium carbonate and calcium carbonate. Many hypotheses — some quite bizarre — were proposed to explain how it must have formed. Only in 2013 was this mineral laboratory synthe-sized by a reasonable pathway [23]. The stability of this mineral can be accounted for by the slightly different sizes of cation sites, for the related mineral *ankerite* has a composi-tion: $Ca(Fe(II),Mg,Mn(II))(CO_3)_2$ where the other ions will only substitute for the magnesium ion, not the calcium ion.

Biological Roles of Strontium and Barium

It is rarely realized that Group 2 provides the greatest num-ber of elements with biological roles: magnesium, calcium, strontium, and barium. The roles of magnesium and cal-cium are well-documented, thus, the focus here will be on strontium and barium. Some algae selectively concentrate these ions to form crystals of barium sulfate and strontium sulfate [24]. However, what is of crucial importance is the incorporation of strontium ion into human bone, *hydroxoap-atite*, $Ca_5(PO_4)_3(OH)$. Bone formation favors incorporation of strontium over calcium by a very large factor. Presumably the larger strontium ion (132 pm) fits "more snugly" in the crystal lattice than the smaller calcium ion (114 pm).

Perhaps if the concentration of strontium had been much higher in the geological past, vertebrates might have normally had strontium-containing bones. In the second half of the 20th century, incorporation of radioactive strontium-90 from weapons tests was feared. Now, the addition of (natural) strontium ion to diet and the incorporation into bone is being proposed as a means of combating osteoporosis [25].

Group 13 (Triels)

Is Group 13 really a group? At the top is boron whose chemistry is dominated by unique cluster species. Then comes aluminum, which would be happier in Group 3 (see Chapter 9). Next is gallium with its near room temperature melting point. And at the bottom, there is thallium that likes to masquerade as a Group 11 element or as a lower Group 1 element (see Chapter 10).

Boron Is Not Boring, It's Unique

As soon as a chemist sees an icosohedron, boron comes immediately to mind. This beautiful and symmetrical molecule is the centerpiece of what makes this element unique. Of course, now a plethora of open- and closed cluster molecules and ions are known. Initially, these other species were believed to be simply fragments of an icosahedron. It was in 1971 that Wade showed that this was not true: instead, they were arranged into three families (*closo-*, *nido-*, and *arachno-*). Subsequently refined by Mingos, the criteria for these skeleta are now known as the *Wade–Mingos rules* [26]. The rules provide a straightforward and elegant rationalization of the shapes of "electron-deficient" cluster compounds in terms of the number of skeletal electron pairs (SEPs) these molecules.

Group 14 (Tetrels)

Just as Groups 1, 17, and 18 are regarded as epitomizing the smooth change in properties descending the respective group, Groups 14 to 16 represent the "discontinuity" groups. These are the groups that span the range of element behavior from nonmetal, through metalloid, to metal. In these cases, though there are sometimes similarities in chemical formula of compounds, there is little that can be chosen to select for group trend. In fact, group individuality is more interesting.

Graphite: The Forgotten Allotrope

Though the diamond and fullerene allotropes of carbon have taken the limelight in recent years, here the focus will be on oft-forgotten graphite. Graphite, with its layer structure of conjugated aromatically bonded atoms, has the ability to trap other atoms and molecules between the carbon sheets [27]. These are known as intercalation compounds:

> *Graphite intercalation compounds* (GICs) are complex materials having a formula CX_m where the ion Xn^+ or X^{n-} is inserted between the oppositely charged carbon layers. Typically, m is much less than 1.

GICs are of interest in providing the electrode framework in battery systems. One specific example, of the many known species, is that between graphite and potassium. Molten potassium is absorbed into the black graphite layers to give a bronze-colored ionic solid with limiting composition of $[K]^+[C_8]^-$ [28].

Cubane: A Whole New Field of Inorganic Chemistry

Over the history of organic chemistry, cubane, C_8H_8, was considered simply a hypothetical molecule. With 90° bond angles, no one thought it could actually be synthesized, that is, until it was in 1964 [29]. Not only was it synthesizable but, when produced, it was a stable molecule. Why mention this in a book that is essentially inorganic chemistry? The *pseudo-cubane* structure is one that permeates cluster inorganic chemistry, and by its name, recognizes the simplest structure from which they are all derived. For example, there are the thallium–oxygen pseudo-cubanes, such as $Tl_4(OCH_3)_4$ [30]. Silicon forms pseudo-cubanes, $Si_8(Si^tBuMe_2)_8$. Phosphorus forms pseudo-cubanes where it alternates with boron, or aluminum, or nitrogen, or carbon, such as $P_4(C^tBu)_4$. But of all the pseudo-cubanes, one must take top billing: that of the iron–sulfur pseudo-cubanes that are such crucial redox systems in so many biochemical processes (Figure 7.3) [31].

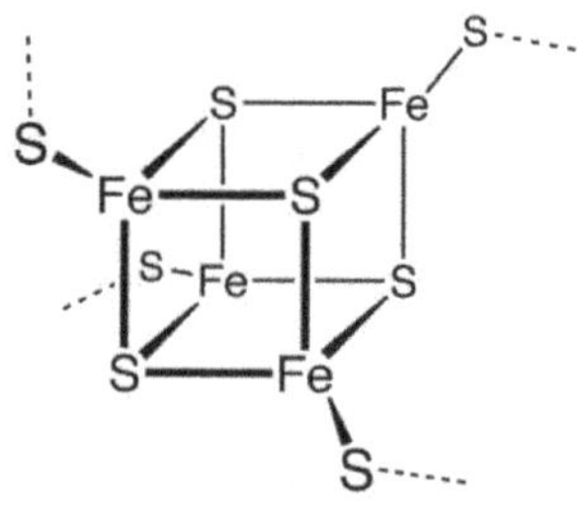

Figure 7.3 The common iron–sulfur pseudo-cubane core of many biological redox systems.

Group 15 (Pnictogens)

As for Group 14, the elements of Group 15 span a wide range of behaviors. And there are always surprises awaiting discovery. As an example, nitrogen is cited as having a single allotrope, N_2. However, we see things from the perspective of our own SATP world. Under the conditions of the very low pressure at the edge of the Earth's atmosphere, tetranitrogen, N_4, is to be found [32].

Tetrahedral Ions

No, not those containing a tetrahedral bonding arrangement, but those species containing a tetrahedron of atoms. The yellow (not white [33]) allotrope of phosphorus, P_4, provides the prototypical example. But just as cubane spawned pseudo-cubanes, so there are other molecules and ions adopting this very bond-strained tetrahedral shape. Some examples of these valence-isoelectronic ions are $[Bi_2Sn_2]^{2-}$, $[Sb_2Pb_2]^{2-}$, $[Si_4]^{4-}$, $[Ge_4]^{4-}$, $[Sn_4]^{4-}$, $[Pb_4]^{4-}$, and $[Tl_4]^{8-}$. Such species are examples of *Zintl ions* [34]. Zintl compounds are brittle, high-melting, intermetallic compounds, which contain polyatomic anions. First investigated in the 1930s, Zintl phases are formed by reacting a Group 1 or Group 2 metal with an element in any of Groups 13, 14, 15, or 16.

Arsenic in Biological Systems

One aspect often overlooked by inorganic chemists is that of substitution of one element for another in a biological organism or process. In this chapter, the substitution of one element by another in the same group will be the focus.

For example, one bacterium can utilize arsenates instead of phosphates [35]. There has even been a computational study of whether such a substitution may be more widespread among bacteria [36].

Group 16 (Chalcogens)

With Group 16, there is the progression from nonmetals, oxygen and sulfur (not "sulphur" [37]); to metalloid, selenium; then to the two weak metals, tellurium and (radioactive) polonium. The allotropes of oxygen used to be dismissed as simply dioxygen and trioxygen (ozone). But no more. Tetraoxygen, O_4, exists at very low pressures in the upper atmosphere [38]; while octaoxygen, O_8, is formed as a dark red solid cubic structure under high pressure, low temperature [39].

Sulfur on Io

Sulfur has more allotropes than any other element [40] and even when it melts upon heating the chemistry continues to be complex [41]. Yet the most interesting sulfur chemistry is not on Earth, but on Jupiter's moon, Io. Most people have seen NASA photos of the startlingly multi-colored moon, unique in the Solar System, which even possesses a sulfur lake [42]. One molecule identified in the atmosphere is disulfur oxide, S_2O [43]. The Io surface colors — yellows, reds, blues, and greens — are almost certainly sulfur species. But what are they? Two possible contributors under the surface bombardment by particles from Jupiter's radiation field are the cations: $[S_8]^{2+}$ (blue) and $[S_{16}]^{2+}$ (red) [44], but no one knows for sure at the date of writing.

Selenium and Tellurium in Biological Systems

Though sulfur-containing amino acids, such as cysteine, are well known, it is little appreciated that the next lower member of the Group, selenium, is also incorporated into biological systems. In fact, seleno-amino acids, specifically seleno-cysteine and seleno-methionine, have essential roles in living organisms, different than those of the sulfur analogues [45]. A phenomenon first observed in 1880, some plants even have a dependence on selenium-rich soils [46]. Descending Group 16 involves crossing from nonmetal to metalloid (see Chapter 5). Despite this change in element properties, there are indeed telluro-cysteine and telluro-methionine [47]. In fact, a certain fungus fed tellurium rather than sulfur quite happily synthesizes the tellurium analogues. Interestingly, a study has compared and contrasted the properties of seleno-cystine and telluro-cystine and their derivatives [48].

Group 17 (Halogens)

After the alkali metals, the halogens provide the second of the Groups that have a unitary classification, this time, of the nonmetals. These highly reactive nonmetals span the range from gases (pale yellow for fluorine, pale green for chlorine) to liquid (almost black, "oily" liquid with red-brown vapor for bromine) to solid (purple-black, metallic-looking solid that melts to a deep violet liquid then boils to a violet gas for iodine).

Weakness of the Fluorine–Fluorine Bond

There is one specific Group 17 feature that affects chemistry throughout the Periodic Table: the weakness of the F–F

Table 7.3 Bond energies for the dihalogen molecules

Dihalogen Molecule	Bond Energy $(kJ \cdot mol^{-1})$
F–F	155
Cl–Cl	240
Br–Br	190
I–I	149

bond (see Table 7.3). The weakness of the F–F bond has to be contrasted with the strength of the (highly polar) bonds of fluorine to other elements. As an example, the F–C bond can be as strong as 544 $kJ \cdot mol^{-1}$. This bond energy difference provides a key factor in the ready formation of fluoro-compounds, particularly those in high oxidation states.

Interhalogen Compounds

If there is one feature unique to the halogens, it must be the readiness to form interhalogen compounds (and inter-pseudo-halogen compounds, see Chapter 14). An *interhalogen compound* contains two different halogen atoms and no atoms of elements from any other group. The formulas are XY_n, where $n = 1, 3, 5,$ or 7, and X is the less electronegative of the two halogens. For $n = 7$, only iodine heptafluoride is known [49]. The halogen pairs also form polyatomic cations and anions that are favorite species for quizzes in general chemistry courses in which molecular geometries have to be deduced using Gillespie–Nyholm (VSEPR) Theory [50]. One fascinating aspect of the interhalogen compounds is the intermediate physical properties to those of the constituent halogens. As examples, chlorine monobromide is a yellow-brown gas at room temperature, while bromine trifluoride is a yellow-green liquid.

Group 18 (Aerogens)

The aerogens/noble gases (Group 18) used to be hailed as exemplars of periodicity with the systematic trends in melting and boiling points. But no more. There seems to be nothing systematic — no "patterns and trends" — about their chemical properties.

Some Xenon Compounds

It is still a common belief that aerogen chemistry is limited to bonds with fluorine or oxygen. Here the focus will be on compounds with other elements, to make the point of the now known diversity of aerogen — particularly xenon — chemistry. Xenon is still by far the most chemistry-rich member of the Group [51]. To stabilize bonds between xenon and less electronegative elements, electron-withdrawing groups on the bonded species are required. These species are most usually fluorine substituted [52]. The prototypical example is the pentafluorophenylxenon(II), $[(C_6F_5)Xe]^+$, ion [53]. This cation is prepared similarly to that of $[(C_6F_5)I]^{2+}$. The electron-withdrawing power of the pentafluorophenyl group is so strong that even chloride ion can be induced to complete the two coordination to form C_6F_5XeCl [54]. However, the most interesting ion including xenon has to be its coordination as a ligand to gold, $[AuXe_4]^{2+}$ [55] (previously mentioned in Chapter 5).

A Selection of Compounds of Other Aerogens

There seem to be some similarities with krypton in that krypton forms an analogous species to xenon, that of C_6F_5KrCl [56]. However, at low temperatures, argon seems to form some unique compounds, such as the now well-established HArF [57]. For highly radioactive radon, there is currently only RnF_2 and RnO_3, which illustrates the common feature that oxides can often be in a higher oxidation state than fluorine [58]. No "real" compounds of helium and neon have been synthesized to the date of writing.

Periodic Trends

The systematic progressions of formulas of hydrides and oxides across each Period was one of the crucial factors in Mendeléev's development of the Periodic Table (see Figure 7.4).

Such a progression is still a fundamental basis of why the Periodic Table is still important today. The definition of trends across a period is best stated as:

Periodic properties are those systematic patterns observed across a Period. Such patterns are commonly trends in chemical formulas of the compounds formed by the elements.

Reihen	Gruppe I. R^2O	Gruppe II. RO	Gruppe III. R^2O^3	Gruppe IV. RH^4 RO^2	Gruppe V. RH^3 R^2O^5	Gruppe VI. RH^2 RO^3	Gruppe VII. RH R^2O^7	Gruppe VIII. — RO^4
1	H=1							
2	Li=7	Be=9.4	B=11	C=12	N=14	O=16	F=19	
3	Na=23	Mg=24	Al=27.3	Si=28	P=31	S=32	Cl=35.5	
4	K=39	Ca=40	—=44	Ti=48	V=51	Cr=52	Mn=55	Fe=56, Co=59, Ni=59, Cu=63.

Figure 7.4 The top part of a Periodic Table published by Mendeléev in 1871 (note that superscripts, not subscripts, were used to identify atom ratios).

Bonding Trends in Main Group Highest Oxidation-State Oxides

It seems therefore appropriate to conclude this chapter with a more detailed examination of periodicity in the formulas of the main group oxides. Table 7.4, as did Mendeléev's table, only displays the highest oxidation-state oxides. The formulas of the highest (oxidation state) oxides correlate with the group number of the nonoxygen element; that is, +1 (Group 1), +2 (Group 2), +3 (Group 13), +4 (Group 14), +5 (Group 15), +6 (Group 16), and +7 (Group 17). The one exception is oxygen difluoride, the only oxide in which the other element has a higher electronegativity than oxygen.

Though there is a smooth progression in formulas (except for the halides), this hides sudden breaks in phase and behavior at room temperature. These changes can be related to the progression of bonding types from ionic to covalent, with the network covalent region marking the borderline between the two bonding categories [59].

The location of the network covalent species shifts diagonally from the 2nd to the 3rd Period. The common explanation is that the location is reflective of the electronegativity, which itself crosses the Periodic Table on a diagonal. The corresponding network covalent molecule in the 4th Period is beneath that in the 3rd, perhaps a reflection of the similarity resulting from the d-block contraction.

As is not uncommon in inorganic chemistry, things do not always fit neat patterns. Specifically, at room temperature, "N_2O_5" has an ionic structure: $[NO_2]^+[NO_3]^-$. The two lower members of Group 16 do not form simple XO_3 molecules, instead, "sulfur trioxide" is a trimer, S_3O_9, containing

Table 7.4 Bonding categories for the 2nd, 3rd, and 4th Period highest oxidation-state oxides

	Group 1	Group 2	Group 13	Group 14	Group 15	Group 16	Group 17
2nd Period	Li_2O	BeO	B_2O_3	CO_2	N_2O_5		F_2O
Bonding type (phase)	Ionic (solid)	Ionic (solid)	Network covalent (solid)	Intermolecular (gas)	Ionic (solid)		Intermolecular (gas)
3rd Period	Na_2O	MgO	Al_2O_3	SiO_2	P_4O_{10}	S_3O_9	Cl_2O_7
Bonding type (phase)	Ionic (solid)	Ionic (solid)	Ionic (solid)	Network covalent (solid)	Intermolecular (gas)	Intermolecular (solid)	Intermolecular (liquid)
4th Period	K_2O	CaO	Ga_2O_3	GeO_2	As_4O_{10}	Se_4O_{12}	Br_2O_5
Bonding type (phase)	Ionic (solid)	Ionic (solid)	Ionic (solid)	Network covalent (solid)	Intermolecular (gas)	Intermolecular (solid)	Intermolecular (solid)

alternating sulfur and oxygen atoms to form a six-membered ring while "selenium trioxide" is an analogous tetramer, Se_4O_{12}. As another "breakdown" of pure periodicity, though chlorine forms a heptaoxide, bromine only forms a pentaoxide, a fact which might be ascribed to the 4th Period anomaly.

Commentary

Periodic patterns and trends are the fundamental basis of the Periodic Table. And it is not all about Groups and Periods, as Rogers has pointed out [60]. However, in this Author's view, sometimes periodicity is raised to almost mythical status in which patterns and trends are "cherry-picked" to illustrate near-perfect sequences. As shown in this chapter, there are many species that "stubbornly" refuse to fit how they "should." As chemists, we should not be afraid to teach the limitations of periodicity and sometimes revel in the uniqueness of each element and its compounds.

References

1. D. G. Cooper, *Chemical Periodicity*, John Murray, London (1974).
2. R. J. Puddephat and P. K. Monaghan, *The Periodic Table of Elements*, 2nd ed., Oxford University Press, Oxford (1986).
3. N. C. Norman, *Periodicity and the p-Block Elements*, Oxford University Press, Oxford (1994).
4. J. Barrett, *Atomic Structure and Periodicity*, Royal Society of Chemistry, London (2002).
5. W. B. Jensen, "The Place of Zinc, Cadmium, and Mercury in the Periodic Table," *J. Chem. Educ.* **80**, 952–961 (2003).

6. M. Niaza, M. Rodrígueza, and A. Brito, "An Appraisal of Mendeleev's Contribution to the Development of the Periodic Table," *Stud. Hist. Phil. Sci. Part A* **34**(2), 271–282 (2004).

7. J. W. van Spronsen, "The Priority Conflict between Mendeleev and Meyer," *J. Chem. Educ.* **46**(3), 136–139 (1969).

8. G. E. Rogers, *Descriptive Inorganic, Coordination, and Solid-State Chemistry*, 3rd ed., Brooks/Cole, 237–239 (2012).

9. M. Kaupp, "Chemical Bonding of Main Group Elements," in G. Frenking and S. Shaik (eds.), *The Chemical Bond: Chemical Bonding across the Periodic Table*, Wiley-VCH, Weinheim, Germany, 4 (2014).

10. P. P. Power, "Main-Group Elements as Transition Metals," *Nature* **463**, 171–177 (2010).

11. J. E. Huuhey and C. L. Huheey, "Anomalous Properties of Elements that Follow 'Long Periods' of Elements," *J. Chem. Educ.* **49**(4), 227–229 (1972).

12. P. Pyykkö, "Relativistic Effects in Structural Chemistry," *Chem. Revs.* **88**(3), 563–594 (1988).

13. J. H. Hildebrand, "The Alternations in Stability of Compounds of the Elements in Group V," *J. Chem. Educ.* **18**(6), 291–292 (1941).

14. W. E. Dasent, "Non-Existent Compounds," *J. Chem. Educ.* **40**(3), 130–134 (1963).

15. R. T. Sanderson, "An Explanation of Chemical Variations within Periodic Major Groups," *J. Am. Chem. Soc.* **74**(19), 4792–4794 (1952).

16. W. E. Dasent, "Textbook Errors, XIV: Arsenic(V) Chloride," *J. Chem. Educ.* **34**(11), 535–536 (1957).

17. A. Y. Herell and K. H. Gayer, "The Elusive Perbromates," *J. Chem. Educ.* **49**(9), 583–586 (1972).

18. D. A. Davenport, "Provocative Opinion: The Grim Silence of Facts," *J. Chem. Educ.* **47**(4), 271 (1970).

19. E. W. Zmaczynski, "Periodic System of the Elements in a New Form," *J. Chem. Educ.* **14**(5), 232–235 (1937).

20. A. Smith, *Introduction to Inorganic Chemistry*, 3rd ed., The Century Company (1917).

21. G. Restrepo, "The Periodic System: A Mathematical Approach," in E. Scerri and G. Restrepo (eds.), *Mendeleev to*

Oganesson: A Multidisciplinary Perspective on the Periodic Table, Oxford University Press, Oxford, 80–103 (2018).

22. Z.-H. Yang, "The Size and Structure of Selected Hydrated Ions and Implications for Ion Channel Selectivity," *RSC Adv.* **5**, 1213–1219 (2015).

23. J. A. Roberts *et al.*, "Surface Chemistry Allows for Abiotic Precipitation of Dolomite at Low Temperature," *Proc. Natl. Acad. Sci.* **110**(36), 1450–14545 (2013).

24. M. R. Krejci *et al.*, "Selectivity in Biomineralization of Barium and Strontium," *J. Struct. Biol.* **176**(2), 192–202 (2011).

25. W. Querido, A. L. Rossi, and M. Farina, "The Effects of Strontium on Bone Mineral: A Review on Current Knowledge and Microanalytical Approaches," *Micron* **80**, 122–134 (2011).

26. A. J. Welch, "The Significance and Impact of Wade's Rules," *Chem. Commun.* **49**, 3615–3616 (2013).

27. M. T. Wheller, *Inorganic Materials Chemistry*, Oxford Science Publications, Oxford University Press, Oxford, 34–35, 69–70 (1994).

28. D. M. Ottmers and H. F.Rase, "Potassium Graphites Prepared by Mixed-Reaction Technique," *Carbon* **4**(1), 125–127 (1966).

29. K. F. Biegasiewicz *et al.*, "Cubane: 50 Years Later," *Chem. Revs.* **115**(14), 6719–6745 (2015).

30. C. E. Housecroft, *Cluster Molecules of the p-Block Elements*, Oxford Science Publications, Oxford, 13–24, 26, 28, 30–33 (1994).

31. M. Ye *et al.*, "A Synthetic Model of Enzymatic [Fe$_4$S$_4$]–Alkyl Intermediates," *J. Am. Chem. Soc.* **141**(34), 13330–13335 (2019).

32. F. Cacace, "From N$_2$ and O$_2$ to N$_4$ and O$_4$: Pneumatic Chemistry in the 21st Century," *Chem. Eur. J.* **8**(17), 3839–3847 (2002).

33. G. Rayner-Canham and A. Young, "What Is the Colour of P$_4$?" *Chem13 News*, 15 (March 2014).

34. S. Scharfe *et al.*, "Zintl Ions, Cage Compounds, and Intermetalloid Clusters of Group 14 and Group 15 Elements," *Angew. Chem. Int. Ed.* **50**(16), 3630–3670 (2011).

35. F. Wolfe-Simon *et al.*, "A Bacterium That Can Grow by Using Arsenic Instead of Phosphorus," *Science* **332**, 1163–1166 (2011).

36. A. K. Jissy and A. Datta, "Can Arsenates Replace Phosphates in Natural Biochemical Processes? A Computational Study," *J. Phys. Chem. B* **117**, 8340–8346.

37. Anon., "Editorial: So Long Sulphur," *Nat. Chem.* **1**, 333 (2009).

38. F. Cacace, G. de Petris, and A. Troiani, "Experimental Detection of Tetraoxygen," *Angew. Chem. Int. Ed.* **40**(21), 4062–4065 (2001).

39. R. Steudel and M. W. Wong, "Dark-Red O_8 Molecules in Solid Oxygen: Rhomboid Clusters, Not S_8-Like Rings," *Angew. Chem. Int. Ed.* **46**(11), 1768–1771 (2007).

40. R. Steudel and B. Eckert, "Solid Sulfur Allotropes," in R. Steudel (Ed.), *Elemental Sulfur and Sulfur-Rich Compounds I, Topics in Current Chemistry 230*, Springer, 1–80 (2003).

41. R. Steudel, "Liquid Sulfur," in R. Steudel (Ed.), *Elemental Sulfur and Sulfur-Rich Compounds I, Topics in Current Chemistry 230*, Springer, 81–116 (2003).

42. J. I. Lunine and D. J. Stevenson, "Physics and Chemistry of Sulfur Lakes on Io," *Icarus* **64**(3), 345–367 (1985).

43. M. Y. Zolotov and B. Fegley Jr., "Volcanic Origin of Disulfur Monoxide (S_2O) on Io," *Icarus* **133**(2), 293–297 (1998).

44. R. J. Gillespie *et al.*, "Polyatomic Cations of Sulfur. I. Preparation and Properties of S_{16}^{2+}, S_8^{2+}, and S_4^{2+}," *Inorg. Chem.* **10**(7), 1327–1332 (1971).

45. A. Böck *et al.*, "Selenocysteine: The 21st Amino Acid," *Mol. Microbiol.* **5**(3), 515–520 (1991).

46. M. Sarquis and C. D. Mickey, "Selenium, Part 1: Its Chemistry and Occurrence," *J. Chem. Educ.* **57**(12), 886–889 (1980).

47. S. E. Ramadan *et al.*, "Incorporation of Tellurium into Amino Acids and Proteins in a Tellurium-Tolerant Fungi," *Biol. Trace Element Res.* **20**(3), 225–232 (1989).

48. K. Satheeshkumar *et al.*, "Reactivity of Selenocystine and Tellurocystine: Structure and Antioxidant Activity of the Derivatives," *Chem. Eur. J.* **24**(66), 17513–17522 (2018).

49. E. H. Wiebenga, E. E. Havinga, and K. H. Boswijk, "Structures of Interhalogen Compounds and Polyhalides," *Adv. Inorg. Chem. Radiochem.* **3**, 133–169 (1961).

50. L. S. Bartell, "A Structural Chemist's Entanglement with Gillespie's Theories of Molecular Geometry," *Coordination Chem. Rev.* **197**(1), 37–49 (2000).

51. S. S. Nabiev, V. B. Sokolov, and B. B. Cahivanov, "Structure of Simple and Complex Noble Gas Fluorides," *Crystallogr. Rep.* **56**(5), 774–791 (2011).

52. W. Henderson, *Tutorial Chemistry Texts, 3: Main Group Chemistry,* Royal Society of Chemistry, Cambridge, 154 (2000).

53. H.-J. Frohn and S. Jakobs, "The Pentafluorophenylxenon(II) Cation: $[C_6F_5Xe]^+$; the First Stable System with a Xenon–Carbon Bond," *J. Chem. Soc. Chem. Commun.* 625–627 (1989).

54. H.-J. Frohn, T. Schroer, and G. Henkel, "C_6F_5XeCl and $[(C_6F_5Xe)_2Cl][AsF_6]$: The First Isolated and Unambiguously Characterized Xenon(II) Chlorine Compounds," *Angew. Chem. Int. Ed.* **38**(17), 2554–2556 (1999).

55. S. Seidel and K. Seppelt, "Xenon as a Complex Ligand: The Tetra Xenono Gold(II) Cation in $AuXe_4^{2+}(Sb_2F_{11}^-)_2$," *Science* **290**(5489), 117–118 (2000).

56. G. J. Schrobilgen, "The Fluoro(hydrocyano)krypton(II) Cation $[HC\equiv N–Kr–F]^+$; the First Example of a Krypton–Nitrogen Bond," *J. Chem. Soc. Chem. Commun.* 863–865 (1988).

57. L. Khriachtchev *et al.* "A Stable Argon Compound," *Nature* **406**, 874–876 (2000).

58. S. Riedel and M. Kaupp, "The Highest Oxidation States of the Transition Metal Elements," *Coordination Chem. Rev.* **253**(5–6), 606–624 (2009).

59. J. Šima, "Structure-Related Melting Points and Boiling Points of Inorganic Compounds," *Found. Chem.* **18**, 67–79 (2016).

60. G. E. Rogers, "A Visually Attractive 'Interconnected Network of Ideas' for Organizing the Teaching and Learning of Descriptive Inorganic Chemistry," *J. Chem. Educ.* **91**, 216–224 (2014).

Chapter 8

Patterns among the Transition Metals

In this chapter, we will deconstruct the monolithic block of transition metals. Traditionally, classification has been by Group, but there are far richer patterns and trends to be found. By using chemical criteria for assignment, a hybrid approach to categorizing and clustering transition metals offers many advantages. Nevertheless, ambiguities arise as to the assignments that we choose.

The first task is to define the transition metals, which is not quite as obvious as it may seem. The main group elements are always identified as the members of Groups 1 to 2, and 13 to 18; that is, the s-block and p-block elements. In addition, about half of textbook sources include Group 12 as main group elements [1]. One might justifiably conclude by deduction that Groups 3 to 11, or Groups 3 to 12, would be the transition metals. As with much of inorganic chemistry, it is not that simple.

What Is a Transition Metal?

The terms "d-block metal" and "transition metal" are not synonymous. Identifying a "transition metal" is not simply a question of location in the Periodic Table, but also one of chemical behavior. A definition common among inorganic chemists is a transition metal is an element that has at least one simple ion with an incomplete outer set of d electrons.

Exclusion of Group 3

Using the criterion earlier, the Group 3 metals are excluded as their common chemistry is all based on the d^0 3+ ion, Sc^{3+} and Y^{3+}. In fact, the chemistry of these two metals more closely resembles that of the lanthanoids. Patterns among the Group 3 elements, therefore, will be discussed in Chapter 12. Of note, the classic series by Sneed and Brasted, *Comprehensive Inorganic Chemistry*, combined scandium and yttrium with the lanthanoids [2].

Exclusion of Group 12

The Group 12 metals are also excluded. For them, the predominant ions, Zn^{2+}, Cd^{2+}, and Hg^{2+} have d^{10} configurations. The isolation of d^8 mercury(IV) fluoride, HgF_4, at very low temperatures [3], initially provoked the claim that mercury should be redesignated as a transition metal. As ever more exotic and fragile species are identified [4], there is the potential for a never-ending expansion of claimed members of the transition metal series.

However, such an eventuality can be avoided by using a definition of:

*A **transition metal** is an element that has at least one simple ion with an incomplete outer set of d electrons, which is stable under ambient conditions.*

Here, this definition will be utilized, though it will be shown that there can be considered one exception to the rule. Before doing so, it is appropriate to review the previous categorizations of transition metals.

Exclusion of Honorary Transition Metals

A new term entering the vocabulary of inorganic chemistry is that of *honorary d elements* [5] or *honorary transition metals* [6]. These terms have been devised to describe organometallic compounds of Group 1 or Group 2 elements that, it is claimed, are using their inner d-orbitals in bonding. Such compounds have been identified by computational studies and/or by synthesis under extremely low temperatures. As such, they are excluded by the earlier definition from study here.

Previous Classifications of Transition Metals

One of the modern authoritative works on inorganic chemistry, Greenwood and Earnshaw's *Chemistry of the Elements* [7] treats each of the transition metal groups as individual entities, devoting a chapter to Group 4, one to Group 5, and so on (Figure 8.1).

The species that seem to be Group-specific are the simple carbonyls [8]. The pattern is shown in Table 8.1.

Another common approach is adopted in *Advanced Inorganic Chemistry* (Cotton, Wilkinson, Murillo, and Bochmann) [9], with each of the elements of the first transition series being treated individually, then the 4d–5d pairs of elements being covered in a subsequent section. This arrangement of material is also adopted in *Inorganic Chemistry* (Housecroft and Sharpe) [10] and in *Descriptive Inorganic Chemistry* (Rayner-Canham and Overton) [11].

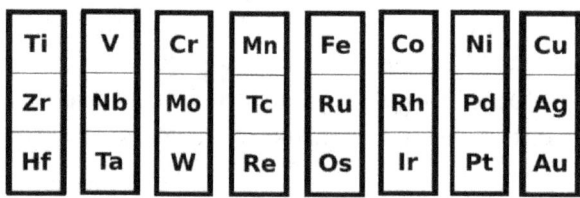

Figure 8.1 The traditional group-by-group study of the transition metals.

Table 8.1 Simple carbonyls of the middle transition metal series

	Group 6	Group 7	Group 8	Group 9
4th Period	$Cr(CO)_6$	$Mn_2(CO)_{10}$	$Fe(CO)_5$	$Co_2(CO)_8$
5th Period	$Mo(CO)_6$	$Tc_2(CO)_{10}$	$Ru(CO)_5$	$Rh_2(CO)_8$
6th Period	$W(CO)_6$	$Re_2(CO)_{10}$	$Os(CO)_5$	$Ir_2(CO)_8$

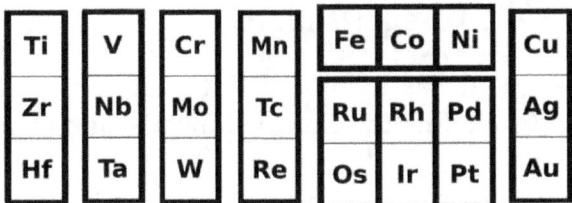

Figure 8.2 Combining the Fe–Co–Ni triad together; and considering the platinum group metals as separate entity.

A third approach is to cover Groups 4 to 7 and 11 individually, then the [Fe–Co–Ni] ferromagnetic triad together, and the platinum metals as a separate entity (Figure 8.2). This format was adopted, among others, by Partington's *General and Inorganic Chemistry for University Students* [12], the series *Pergamon Texts in Inorganic Chemistry* [13, 14], and by *Textbook of Inorganic Chemistry* (Gopalan) [15]. Historically, the platinum group metals, since their discovery, have always been considered as a related "cluster" [16]. Lee has commented that, for the Groups 8, 9, and 10 [17]:

> . . . *the horizontal similarities between these elements are greater than anywhere in the periodic table except among the lanthanides.*

He also noted that the similarities:

> . . . *are sometimes emphasized by considering these nine elements as two horizontal groups: the three ferrous metals Fe, Co and Ni, and the six platinum metals Ru, Rh, Pd, Os, Ir and Pt.*

Each of these classification systems has one flaw — that they organize the transition metals largely according to one strategy and they define the trends according to that organization. Thus linkages, relationships, patterns, or similarities outside of that framework are ignored. Two exceptions have been the proposals by Habashi and by Schweitzer and Pesterfield.

Habashi's Categorizations

Habashi [18] has identified three categories of transition metals (excluding the Group 11 metals) and named them as follows:

- The *vertical similarity transition metals*: [Zr–Hf]; [Nb–Ta]; [Mo–W]; and [Tc–Re].
- The *horizontal similarity transition metals*: [Ti–V–Cr–Mn–Fe–Co–Ni].
- The *horizontal–vertical transition metals*: [Ru–Os–Rh–Ir–Pd–Pt].

That is, Habashi considered the 3d metals as separate entities from the 4d and 5d metals (Figure 8.3). There is much chemical evidence for the 4d and 5d metals, as a set, being very different to those of the 3d metals. In particular, the chemistry of zirconium and hafnium is almost identical, yet significantly different to that of titanium [19].

One of the vertical similarity transition sets has a biochemical basis. Certain bacteria, which normally utilize molybdenum in some of their enzymes, utilize tungsten instead when the bacteria are in high-temperature environments. It is believed that the tungsten-containing enzymes can survive and function as thermophiles [20].

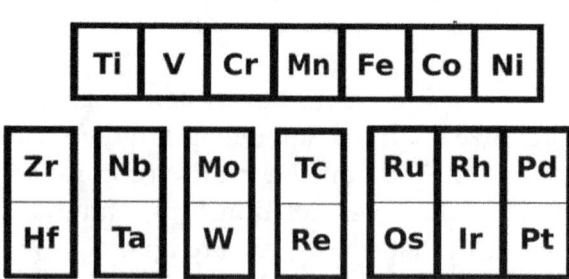

Figure 8.3 The transition metal classification according to Habashi [18].

Schweitzer and Pesterfield's Categorizations

The reference work, the *Aqueous Chemistry of the Elements* (Schweitzer and Pesterfield), includes a series of Pourbaix diagrams [21]. The authors assign the 3d transition metals to two horizontal triads: [V–Cr–Mn] that form compounds in the maximum oxidation states; and [Fe–Co–Ni] for which +2 and +3 oxidation states predominate [22]. Schweitzer and Pesterfield treat copper with the other two Group 11 metals as the [Cu–Ag–Au] vertical triad, while titanium is placed in a chapter with all the 4d and 5d transition metals. They divide the heavy transition metals (plus titanium) into three subcategories:

- The elements for which insoluble oxides dominate [Ti–Zr–Hf–Nb–Ta].
- The elements with high oxidation-state oxo-anions [Mo–W–Tc–Re].
- The platinum metals [Ru–Os–Rh–Ir–Pd–Pt].

This scheme is shown in Figure 8.4.

Other Categorizations

In recent years, using chemotopological methods, there have been new attempts at classifications of the elements. Sneath's study [23] divided the heavy transition metals into

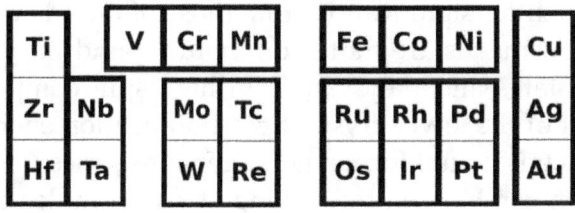

Figure 8.4 The transition metal classification according to Schweitzer and Pesterfield [21].

one cluster and the later ones (plus gold) into a separate cluster. The analysis by Leal *et al.* [24] suggested that, for the 3d metals, there was a [Cr–Fe–Co–Ni] linkage, while titanium belonged to Group 4 as [Ti–Zr–Hf] and manganese and vanadium were unique in their chemistry. For the 4d–5d transition metals, they proposed the following linkages: [Zr–Hf (with Ti)]; [Mo–W (with Ge)]; [Nb–Ta–Tc–Re]; [Ru–Os]; [Rh–Ir–Pd–Pt]; and [Ag–Au].

Categorizations of the Transition Metals

In this chapter, a schema will be deduced from first principles using chemical criteria that have similarities to, but also differences from, those of Habashi and of Schweitzer and Pesterfield. Though copper and gold fit the chemical transition metal criteria, silver does not. In developing these criteria, flexibility in any classification is necessary and, indeed, at least two elements might be considered as having "secondary allegiances."

The 3d Period Patterns

Just as the 2nd Period main group elements differ from those of the subsequent Periods (see Chapter 7), it has always been recognized that the metals of the 3d transition series differ significantly from those of the 4d and 5d series. The 3d metals are more commonly found in lower oxidation states and they can form high-spin compounds as a result of the lower crystal field stabilization energy. But the 3d metals do not form a homogeneous series. This differentiation can be seen by looking at the formulas of the common oxides (Table 8.2) [25].

Table 8.2   The common oxides of the 3d transition metals

Ox. No.	Titanium	Vanadium	Chromium	Manganese	Iron	Cobalt	Nickel	Copper
+7				Mn_2O_7				
+6			CrO_3					
+5		V_2O_5						
+4	TiO_2	VO_2	CrO_2	MnO_2			NiO_2	
+3	Ti_2O_3	V_2O_3	Cr_2O_3	Mn_2O_3	Fe_2O_3			
+2 and +3				Mn_3O_4	Fe_3O_4	Co_3O_4	Ni_3O_4	
+2	TiO	VO		MnO	FeO	CoO	NiO	CuO
+1								Cu_2O

Among the many oxides, for the Group 4 to Group 7 metals, there is an oxide series for which the metal has its maximum oxidation state (TiO_2, V_2O_5, CrO_3, and Mn_2O_7). For the same set of metals, there is also an oxide series MO_2 (with M = Ti, V, Cr, and Mn). On the basis of the oxides, it could be considered that there is a [Ti–V–Cr–Mn] tetrad subgroup of the 3d metals. Likewise, the next three of the 3d transition metals, the [Fe–Co–Ni] triad, are characterized by having +2 and +3 oxidation states in their common oxides. Copper is unique among the 3d metals in exhibiting an oxidation state of +1 in an oxide (and in its chemistry in general). Of course, the divisions are not clear-cut. For example, there is the oxide series of M_3O_4 that encompasses manganese through to nickel.

The decision of which elements belong in what groupings will always be subjective. This Author places the greater weight on the species under strongly oxidizing conditions. To illustrate this point, Table 8.3 shows that, under strongly oxidizing conditions, manganese completes the set of isoelectronic highly oxidizing anions. Whereas vanadium, chromium, and manganese all form soluble tetroxo-anions, titanium forms an insoluble oxide. For this reason, on balance, titanium is the "weakest link" in this set. In addition, the tetroxo-anions of the [V–Cr–Mn] triad form a series of increasing acid strength.

Table 8.3 Comparative species for the [Ti–V–Cr–Mn] tetrad over the pH range under strongly oxidizing conditions

	Very Acidic	**Acidic**	**Basic**	**Very Basic**
Titanium	$TiO^{2+}(aq)$	$TiO_2(s)$		
Vanadium	$VO_2^+(aq)$	$H_2VO_4^-(aq)$	$HVO_4^{2-}(aq)$	$VO_4^{3-}(aq)$
Chromium	$H_2CrO_4(aq)$	$HCrO_4^-(aq)$	$CrO_4^{2-}(aq)$	
Manganese	$MnO_4^-(aq)$			

The Trouble with Titanium

Without an aqueous chemistry over most of a pH range, titanium does not appear to fit with its 4th Period neighbors of Group 5, 6, and 7. In fact, titanium is a troublesome element in the context of placement. Should it be considered as being the beginning member of the 4th Period transition series or as the top member of Group 4? Table 8.4 shows the similarity over the pH range with the other members of Group 4.

But before deciding this to be the definitive solution, it is necessary to look at the comparative species at very low pH while reducing the potential. Titanium, but not zirconium or hafnium, has a significant chemistry of its +3 and even +2 oxidation states.

In fact, from a redox perspective, titanium chemistry matches more with that of vanadium and less with zirconium. As can be seen in Table 8.5, there is a remarkable similarity in oxidation states and species, taking into consideration that the maximum oxidation state of vanadium is +5 while that for titanium is +4.

Thus, titanium lays claim both to be the top member of the Group 4 triad [Ti–Zr–Hf] and the first member of the early 3d transition series tetrad [Ti–V–Cr–Mn]. On balance, because of the dominance of the insoluble +4 oxidation-state

Table 8.4 Comparative species for the [Ti–Zr–Hf] triad over the pH range under strongly oxidizing conditions

	Very Acidic	Acidic	Basic	Very Basic
Titanium	$TiO^{2+}(aq)$		$TiO_2(s)$	
Zirconium	$ZrO^{2+}(aq)$		$ZrO_2(s)$	
Hafnium	$HfO^{2+}(aq)$		$HfO_2(s)$	

Table 8.5 A comparison of species for the [Ti–Zr–Hf] triad at very low pH over the redox range

	Strongly Reducing	Reducing	Near-Zero Potential	Oxidizing	Strongly Oxidizing
Titanium	Ti(s)	Ti^{2+}(aq)	Ti^{3+}(aq)	TiO^{2+}(aq)	
Vanadium	V(s)	V^{2+}(aq)	V^{3+}(aq)	VO^{2+}(aq)	VO$_2^{+}$(aq)
Zirconium	Zr(s)	ZrO^{2+}(aq)			
Hafnium	Hf(s)	HfO^{2+}(aq)			

oxide, the link of titanium with zirconium and hafnium seems to be the stronger.

Manganese Muddies Things

Just as titanium has two allegiances at the beginning of the 3d row, so manganese in the middle also presents a dilemma. Table 8.3 shows that, under highly oxidizing conditions, manganese completes the set of isoelectronic highly oxidizing anions. However, under normal conditions of aqueous chemistry, manganese favors the +2 oxidation state and its species match well with the subsequent members of the 3d series. And, as alluded to earlier, manganese forms Mn_3O_4 — part of the mixed oxidation-state oxide series running from manganese to nickel. So manganese, like titanium, has a "dual identity."

Categorizing the Early 4d–5d Elements

Before dividing up these elements, there is actually at least one isoelectronic series that spans all the 5d elements; that is, the hexacarbonyls [26]. The series is shown in Table 8.6.

Table 8.6   The isoelectronic hexacarbonyl-complexes of the 5d transition elements

	Group 4	Group 5	Group 6	Group 7	Group 8	Group 9
5th Period	$[Hf(CO)_6]^{2-}$	$[Ta(CO)_6]^-$	$[W(CO)_6]$	$[Re(CO)_6]^+$	$[Os(CO)_6]^{2+}$	$[Ir(CO)_6]^{3+}$

Table 8.7   The isostructural octofluoro-complexes of the early 4d–5d transition elements

	Group 4	Group 5	Group 6	Group 7
4th Period	ZrF_8^{4-}	NbF_8^{3-}	MoF_8^{2-}	TcF_8^{2-}
5th Period	HfF_8^{4-}	TaF_8^{3-}	WF_8^{2-}	ReF_8^{2-}

However, there is a fundamental difference between the early 4d–5d elements and the later ones: size. The early heavier transition metal ions are significantly larger, enabling them to have coordination numbers up to eight. An excellent example is the series of isostructural (though not all valence-isoelectronic) octafluoro-complexes as shown in Table 8.7.

On this basis, should [Zr–Hf–Nb–Ta–Mo–W–Tc–Re] be considered an octad of elements? One pattern does not make a cluster, but on the other hand, if we took every single piece of evidence, we would consider each element completely unique.

A defining difference between the [Zr–Hf–Nb–Ta] tetrad and [Mo–W–Tc–Re] tetrad is the difference in aqueous chemistries of the two tetrads. The simple chemistry of the first tetrad across most of the pH range is defined by the insoluble oxides: ZrO_2, HfO_2, Nb_2O_5, and Ta_2O_5. For the second tetrad, it is the soluble tetraoxo-anions that dominate MoO_4^{2-}, WO_4^{2-}, TcO_4^-, and ReO_4^-.

The Platinum Metals

The platinum metal group consists of the [Ru–Os–Rh–Ir–Pd–Pt] hexad of elements. Livingstone has commented that a distinction between the Fe–Co–Ni series and the lower members of the respective Groups is that they form hexahydrated dipositive ion, while the platinum metals do not. However, he cautioned that [27]:

> Because of this difference between iron, cobalt and nickel, on the one hand, and the platinum metals on the other, it must not be overlooked that the relationships within the Group VIII are vertical.

Livingstone provided some examples:

- Iron, ruthenium, and osmium (Group 8) all form carbonyls of formula $M(CO)_5$.
- Cobalt, rhodium, and iridium (Group 9) all form carbonyls of formula $M_2(CO)_8$.
- Nickel, palladium, and platinum (Group 10) all form tetracyano-anions $[M(CN)_4]^{2-}$.

Nevertheless, there is a general acceptance that the six platinum metals form a cluster. Their simple chemistry is characterized by insoluble oxides under strongly oxidizing conditions.

- For [Ru–Os], the +8 oxidation state is favored: RuO_4 and OsO_4.
- For [Rh–Ir–Pd–Pt], the +4 oxidation state is favored: RhO_2, IrO_2, PdO_2, and PtO_2.

As these six elements are found naturally in their elemental state — and often alloyed together — they have been as much an interest of geologists and metallurgists as chemists. Darling has commented [28]:

> In 1860 Claus announced his view that the platinum metals formed "an isolated metallic group, inseparable and solidly constituted." The physical and metallurgical evidence that has since been accumulated fully confirms the validity of this early chemical generalization.

Geochemists and metallurgists subdivide the platinum metals into the iridium–platinum group elements (IPGEs), [Os–Ir–Ru], and the palladium–platinum group elements (PPGEs), [Rh–Pt–Pd]. The distinction within the platinum metal bloc arises from the IPGEs existing almost exclusively in elemental form (siderophiles), while the PPGEs can also be found as metal sulfides (chalcophiles) [29].

Is There, in Fact, a Group 11?

Using the word "Group" implies similarity between members. For example, Group 4 was simple. The oxidation state of +4 dominated all of the three elements. The three elements, copper, silver, and gold of so-called "Group 11," can scarcely be said to form a "group." Thompson has pointed out [30]:

> Silver is the middle element of the Group IB series, yet in many of its physical properties it exhibits extreme values rather than values which fall between those of copper and gold, e.g. melting point (min.), boiling point (min.), thermal conductivity (max.), etc. It exhibits a steel grey colour similar to that of a majority of metals similar to that of the majority of metals rather than a colour similar to that of copper or gold.

To extend the argument to the chemical properties, as shown in Table 8.8: for copper, the +2 oxidation state dominates; for silver, the +1 state; and for gold, the +3 state.

Table 8.8   Comparative species for the [Cu–Ag–Au] ions under oxidizing conditions

	Very Acidic	Acidic	Basic	Very Basic
Copper	$Cu^{2+}(aq)$		$Cu(OH)_2(s)$	$Cu(OH)_4^{2-}(aq)$
Silver	$Ag^+(aq)$		$Ag_2O(s)$	$AgO^-(aq)$
Gold	$Au_2O_3(s)$			$Au(OH)_4^-(aq)$

Table 8.9   Comparative species for the [Fe–Co–Ni–Cu] tetrad under mild oxidizing conditions

	Very Acidic	Acidic	Basic	Very Basic
Iron	$Fe^{2+}(aq)$		$Fe(OH)_2(s)$	
Cobalt	$Co^{2+}(aq)$		$Co(OH)_2(s)$	$Co(OH)_4^{2-}(aq)$
Nickel	$Ni^{2+}(aq)$		$Ni(OH)_2(s)$	
Copper	$Cu^{2+}(aq)$		$Cu(OH)_2(s)$	$Cu(OH)_4^{2-}(aq)$

In fact, in terms of its coordination chemistry, copper fits better with the later 3d elements. For example, in the species across the oxidizing pH range, we see a strong similarity (Table 8.9).

The dominance of the +1 d^{10} state for the normal aqueous chemistry of silver [31] make it more appropriately considered as a main group metal. As to be discussed in Chapter 10, silver is the "classic" example of the "knight's move" linkage, showing a startling similarity to thallium. A unique parallel is that they are the only two metal ions to form brick-red insoluble chromates, Ag_2CrO_4 and Tl_2CrO_4. Under oxidizing conditions, we have the parallel shown in Table 8.10.

Gold is a very different element to that of copper and silver. In fact, it has been referred to as the *gold anomaly* [32]. This anomaly is largely ascribed to the importance of the

Table 8.10 Comparative species for the [Ag–Tl] diad under oxidizing conditions

	Very Acidic	Acidic	Basic	Very Basic
Silver	$Ag^+(aq)$		$Ag_2O(s)$	$AgO^-(aq)$
Thallium	$Tl^+(aq)$			$TlO^-(aq)$

Table 8.11 Comparative species for the [Au–Pt] diad under oxidizing conditions

	Very Acidic	Acidic	Basic	Very Basic
Gold	$Au_2O_3(s)$			$Au(OH)_4^-(aq)$
Platinum	$PtO_2(s)$			$Pt(OH)_6^{2-}(aq)$

relativistic effect, as mentioned in Chapter 2. Just as there are unusual species linking silver(I) and thallium(I), so gold(V) shows a strong resemblance to platinum(V). A good example is the pair of compounds: $[O_2]^+[PtF_6]^-$ and $[O_2]^+[AuF_6]^-$.

The species for platinum and gold can be compared under oxidizing conditions. Allowing for the fact that +4 is the dominant oxidation state of platinum, and six, its common coordination number, there is again a much closer parallel of gold with platinum than with silver and copper (Table 8.11).

Gold is actually the easiest one of this Group to assign to a cluster: that is, to the platinum metals. It is the platinum metals plus gold that are the only metals found almost entirely in their elemental state [33]. In fact, gold is sometimes found in nature as an alloy with platinum and palladium [34]. The acceptance of this "cluster" was cited by Barnes *et al.* [35]:

> *Au is strictly speaking not a platinum-group element but it is a noble metal and will be included with the PGE's in this paper.*

This Author supports the addition of gold to the platinum metals with the designation of "noble metals" (see Chapter 5).

Geochemists often extend this cluster to include rhenium. These eight elements, [Os–Ir–Ru–Rh–Pt–Pd–Re–Au], are referred to as the *highly siderophile elements* (HSE), that is, elements found throughout the solar system most commonly in elemental form [36].

A Hybrid Solution

So, is there a classification of the transition elements that better reflect the linkages? Obviously, we cannot satisfy all the many similarities, but the argument is made here that the best fit is not accomplished by either the Group or the Period approach. Instead, a hybrid combination generates the clusters of elements that have similarities worthiest of highlighting (Figure 8.5). As one example of the allegiance of titanium to the [Zr–Hf–Nb–Ta] tetrad is that all five of the elements form trisulfides of the form MS_3 [37].

Subsequent to deducing the splitting of allegiances for titanium and manganese, Leal and Restrepo have highlighted the same divisions in their "ordered hypograph" [38].

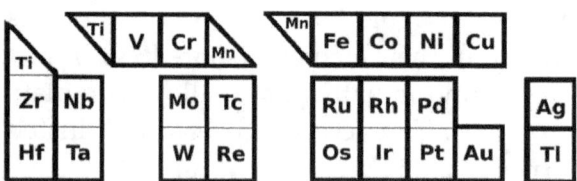

Figure 8.5 A hybrid approach to transition metal classification.

To review, the most logical clustering is listed in the following, showing the "secondary allegiances" of titanium and of manganese:

- The [Ti–Zr–Hf–Nb–Ta] pentad whose simple chemistry is dominated by insoluble oxides.
- The [V–Cr–Mn] triad that exhibits soluble, oxidizing, isoelectronic tetraoxo-anions plus a stable +3 oxidation state, to which Ti can be appended for other aspects of common chemistry.
- The [Fe–Co–Ni–Cu] tetrad for which the +2 aqueous ion is a major component of simple chemistry, to which Mn can be appended for some commonalities.
- The [Mo–W–Tc–Re] tetrad for which nonoxidizing soluble valence-isoelectronic tetraoxo-anions exist.
- The [Ru–Os–Rh–Ir–Pd–Pt–Au] heptad that can be defined as the "noble metals."
- The [Ag–Tl] diad that reminds us to "think outside the (transition metal) box."

Commentary

Chemists like smooth patterns — continuities — systematic trends. Sorry, it doesn't happen with the transition metals! Instead, each of the transition metals flaunts its individuality and refuses to fit neatly into a specific category. The 3d metals are fractured into two halves; zirconium and hafnium behave like twins; silver seems more at home outside the transition metals; gold nestles up to the platinum metals; and titanium and manganese have split allegiances between their Period and their Group. Never let it be said that transition metal chemistry is predictable and boring!

References

1. W. B. Jensen, "The Place of Zinc, Cadmium, and Mercury in the Periodic Table," *J. Chem. Educ.* **80**, 952–961 (2003).
2. M. C. Sneed and R. C. Brasted (Eds.), "Scandium, Yttrium, and the Lanthanide Series," in *Comprehensive Inorganic Chemistry*, vol. 4, D. Van Nostrand Co. Inc., 1955.
3. X. Wang *et al.*, "Mercury Is a Transition Metal: The First Experimental Evidence for HgF_4," *Angew. Chem. Int. Ed.* **46**, 8371–8375 (2007).
4. W. B. Jensen, "Is Mercury Now a Transition Element?" *J. Chem. Educ.* **85**, 1182–1183 (2008).
5. L. Gagliardi and P. Pyykkö, "Cesium and Barium as Honorary *d* Elements: CsN_7Ba as an Example," *Theor. Chem. Acc.* **110**(3), 205–210 (2003).
6. X. Wu *et al.*, "Barium as Honorary Transition Metal in Action: Experimental and Theoretical Study of $Ba(CO)^+$ and $BaCO)^-$," *Angew. Chem. Int. Ed.* **130**(48), 15856–15857 (2018).
7. N. N. Greenwood and A. Earnshaw, *Chemistry of the Elements*, 2nd ed., Butterworth-Heinemann, Oxford, 1997.
8. J. G. Ameen and H. F. Durfee, "The Structure of Metal Carbonyls," *J. Chem. Educ.* **48**(6), 372–375.
9. F. A. Cotton *et al.*, *Advanced Inorganic Chemistry*, 6th ed., Wiley-Interscience, New York, 1999.
10. C. E. Housecraft and A. G. Sharpe, *Inorganic Chemistry*, 4th ed., Pearson, Harlow, 2012.
11. G. Rayner-Canham and T. Overton, *Descriptive Inorganic Chemistry*, 6th ed., W. H. Freeman, New York, 2014.
12. J. R. Partington, *General and Inorganic Chemistry for University Students*, 3rd ed., Macmillan and Co. Ltd., London, 1958.
13. D. Nicholls, *The Chemistry of Iron, Cobalt and Nickel (Pergamon Texts in Inorganic Chemistry, volume 24)*, Pergamon Press, Oxford, 1975.
14. S. E. Livingstone, *The Chemistry of Ruthenium, Rhodium, Palladium, Osmium, Iridium and Platinum (Pergamon Texts in Inorganic Chemistry, volume 25)*, Pergamon Press, Oxford, 1975.

15. R. Gopalan, *Textbook of Inorganic Chemistry*, CRC Press, Boca Raton, 2012.
16. W. P. Griffith, "The Group VIII Platinum-Group Metals and the Periodic Table," *Found. Chem.* **12**, 17–25 (2010).
17. J. D. Lee, *Concise Inorganic Chemistry*, 5th ed., Chapman & Hall, London (1996).
18. F. Habashi, "Metals: Typical and Less Typical, Transition and Inner Transition," *Found. Chem.* **12**, 31–39 (2010).
19. W. C. Fernelius, "Hafnium," *J. Chem. Educ.* **59**(3), 242 (1982).
20. M. J. Pushie and G. N. George, "Spectroscopic Studies of Molybdenum and Tungsten Enzymes," *Coordination Chem. Rev.* **255**(9–10), 1055–1084 (2011).
21. G. K. Schweitzer and L. L. Pesterfield, *Aqueous Chemistry of the Elements*, Oxford University Press, Oxford (2010).
22. D. Venkataraman *et al.*, "A Coordination Geometry Table of the d-Block Elements and Their Ions," *J. Chem. Educ.* **74**(8), 915–918 (1997).
23. P. H. A. Sneath, "Numerical Classification of the Chemical Elements and Its Relation to the Periodic System," *Found. Chem.* **2**, 237–263 (2000).
24. W. Leal, G. Restrepo, and A. Bernal, "A Network Study of Chemical Elements: From Binary Compounds to Chemical Trends," *MATCH Commun. Math. Comput. Chem.* **68**, 417–442 (2012).
25. C. J. Jones, *d- and f-Block Chemistry*, Royal Society of Chemistry, Cambridge, 40 (2001).
26. C. Bach *et al.*, "Cationic Iridium(III) Carbonyl Complexes: $[Ir(CO)_6]^{3+}$ and $[Ir(CO)_5Cl]$," *Angew. Chem. Int. Ed.* **35**(17), 1974–1976 (1996).
27. S. E. Livingstone, *The Chemistry of Ruthenium, Rhodium, Palladium, Osmium, Iridium and Platinum (Pergamon Texts in Inorganic Chemistry, volume 25)*, Pergamon Press, Oxford, 1189 (1973).
28. A. S. Darling, "Some Properties and Applications of the Platinum Group Metals," *Int. Metall. Rev.* **18**, 91–122 (1973).
29. S. K. Mondal, "Platinum Group Element (PGE) Geochemistry to Understand the Chemical Evolution of the Earth's Mantle," *J. Geol. Soc. India* **77**, 295–302 (2011).

30. N. R. Thompson, "Silver," in *The Chemistry of Copper, Silver and Gold (Pergamon Texts in Inorganic Chemistry, volume 17)*, Pergamon Press, Oxford, 83 (1973).

31. J. A. McMillan, "Higher Oxidation States of Silver," *Chem. Rev.*, **62**, 65–80 (1962).

32. P. Schwerdtfeger, "Relativistic Effects in Properties of Gold," *Heteroat. Chem.* **13**, 578–584 (2002).

33. R. Hulme, "Mineral Sources and Extraction Methods for the Elements: A Correlation Based on Elements' Positions in the Periodic Table," *J. Chem. Educ.* 3(3), 111–113 (1956).

34. A. Knopf, "A Gold-Platinum-Palladium Lode in Southern Nevada," *Contrib. Econ. Geol.* 1–19 (1915).

35. S.-J. Barnes, A. J. Naldrett, and M. P. Gorton, "The Origin of the Fractionation of Platinum-Group Elements in Terrestrial Magmas," *Chem. Geol.* **53**, 303–323 (1985).

36. J. M. D. Day, "Highly Siderophile Elements in Earth, Mars, the Moon, and Asteroids," *Rev. Mineral Geochem.* **81**, 161–238 (2016).

37. E. Sandre, A. LeBlanc, and M. Danot, "Giant Molecules in Solid State Chemistry: Using Niobium Trisulfide as an Example," *J. Chem. Educ.* **68**(10), 809–815.

38. W. Leal and G. Restrepo, "Formal Structure of Periodic System of Elements," *Proc. R. Soc. Proc. A* **475**, 20180581 (2019).

Chapter 9

Group (n) and Group ($n + 10$) Relationships

The early Periodic Tables displayed an eight-Group system. Though we now use an 18-Group array, the old versions were based on evidence of similarities between the elements in what we now label the Group (n) and the corresponding elements of Group (n + 10). In this chapter, these similarities will be explored in depth. It is shown that such linkages are not limited to the top members of each Group as earlier discussions have emphasized.

In Chapter 6, the term pseudo-isoelectronic was introduced. This term described a subset of valence-isoelectronic linking a main group element and a transition element. Such electron configurations differ specifically by d^{10} or $f^{14}d^{10}$. For example, the Mg^{2+} ion and the Ca^{2+} ion are valence-isoelectronic, but they each differ from the Zn^{2+} ion by the $3d^{10}$ electrons. Therefore, the Zn^{2+} ion is pseudo-isoelectronic of the two main group ions. This chapter specifically focuses upon these pseudo-isoelectronic relationships.

Going Back to the Past

It was Newlands who first proposed that the chemical elements could be organized according to the "Law of Octaves" [1, 2]. The later Periodic Tables produced by Mendeleev also utilized an eight-column table. This was done even though it meant that the [Fe–Ni–Co]; [Ru–Pd–Rh]; and [Os–Pt–Ir] series were squished into the single Group VIII (see Figure 9.1) [3]. Nevertheless, a key point of the table occupancy was the element similarities within each of the Groups. As an example, in Group V, there were similarities among all eight elements: nitrogen, phosphorus, vanadium, arsenic, niobium, antimony, tantalum, and bismuth.

I	II	III	IV	V	VI	VII	VIII
	H						
Li	Be	B	C	N	O	F	
	Na	Mg	Al	Si	P	S	Cl
K	Ca		Ti	V	Cr	Mn	Fe Co Ni Cu
	(Cu)	Zn			As	Se	Br
Rb	Sr	(Y)	Zr	Nb	Mo		Ru Rh Pd Ag
	(Ag)	Cd	In	Sn	Sb	Te	I
Cs	Ba		Ce				
				Ta	W		Os Ir Pt Au
	(Au)	Hg	Tl	Pb	Bi		
			Th		U		

Figure 9.1 A simplified version of one of Mendeléev's designs of the Periodic Table.

The Rise of the Long Form of the Periodic Table

In 1893, Rang devised one of the first long-form Periodic Tables [4]. He numbered the columns from I through VIII then I through VII (as the noble gases were then unknown). With duplicate numbering for Groups I through VII, it seems to have been Deming in 1923 who first used "A" and "B" designations in a pedagogical context to clarify which groups were which [5]. This was the system adopted by the American Chemical Society (ACS), such that what we now call Group 3 was labeled as "IIIB" while Group 13 was labelled "IIIA." However, the International Union of Pure and Applied Chemistry (IUPAC) adopted a system whereby what we now call Group 3 was labeled as "IIIA" and Group 13 was labeled "IIIB" [6].

It was to resolve this confusion, that the 1 through 18 notations were proposed by IUPAC in the 1980s and subsequently adopted worldwide [7]. A disadvantage has been that the similarities that caused Mendeleev and others to put, for example, silicon and titanium in the same Group, became largely forgotten or overlooked. Only Sanderson in

1954 bravely continued to expound the pedagogical benefits of the eight-column table [8]. For example, in his opinion, that it made much more sense, in teaching general chemistry to have a unitary Group 3 consisting of B–Al–Sc–Ga–Y–In–La–Tl–Ac, all of which have a common oxidation state of +3. The common oxidation states of the other Groups: 0 to VII can be readily identified in a similar manner.

The Rediscovery of the A and B Links

It was Laing who, in 1989, first reminded the modern generations of chemists of these similarities [9, 10]. He noted the resemblances between silicon and titanium compounds (such as the pair $SiCl_4$ and $TiCl_4$); phosphorus and vanadium compounds (such as $POCl_3$ and $VOCl_3$); sulfur and chromium polyatomic ions (such as SO_4^{2-} and CrO_4^{2-}); and chlorine and manganese compounds (such as Cl_2O_7 and Mn_2O_7). To emphasize the linkage, Laing proposed that the element "boxes" of lithium to fluorine and sodium to chlorine be repeated above the corresponding transition metal column.

Rich [11] proposed a modification to Laing's diagram: that oxygen and fluorine be deleted from the duplication as there are no similarities between those elements and the corresponding transition metals of chromium and manganese. In fact, there is little similarity between any of the 2nd Period main group elements and the corresponding d-block elements. It is of note that all of Laing's examples compare 3rd Period main group elements with the 4th Period d-group elements. Thus, the segment of the Periodic Table in the following, derived from Rich and Lang, simply show the addition of the respective Period 3 main group elements to the top of the respective transition metal columns (Figure 9.2).

Laing's study focused on the formula similarities of the 3rd Period main group elements with the corresponding

Group 3	Group 4	Group 5	Group 6	Group 7		Group 12
Al	Si	P	S	Cl		Mg
Sc	Ti	V	Cr	Mn		Zn
Y	Zr	Nb	Mo	Tc		Cd

Figure 9.2 A segment of the Periodic Table showing the proposed additional 3rd Period members of Groups 3–7 and 12.

Groups of the transition series. However, Mingos showed that such parallels in formula existed in other (n) and ($n + 10$) pairs [12]. Here, a wide exploration of such connections will be made.

Definition of the Group (n) and Group ($n + 10$) Relationship

The linkage in chemical formulas and chemical behavior between each Group (n) member and the corresponding Group ($n + 10$) member is quite specific. The relationship is between compounds and polyatomic ions of the highest oxidation state of the main group elements and those of the same oxidation state of the matching transition elements. A general definition is [13]:

> The **(n) and ($n + 10$) relationship** identifies some similarities in some of Group (n) members with those of the corresponding Group ($n + 10$). This resemblance is usually in the highest oxidation state. Such similarities can be in chemical formulas and structures of compounds and polyatomic ions, and of their aqueous behavior.

The (n) and ($n + 10$) linkage for highest oxidation states comes about through electronic structural similarities. That

Table 9.1 Similarities in formula of some oxo-anions to illustrate the (n) and ($n + 10$) relationship

Group	Group 5 and 15	Group 6 and 16	Group 7 and 17
(n) 4th Period	VO_4^{3-}	CrO_4^{2-}	MnO_4^{-}
($n + 10$) 3rd and 4th Period	PO_4^{3-}	SO_4^{2-}	ClO_4^{-}
	AsO_4^{3-}	SeO_4^{2-}	BrO_4^{-}

is, the (n) element in its highest oxidation state has a noble gas electron configuration while the corresponding ($n + 10$) element in its highest oxidation state has, in addition, a filled d^{10} set. For the elements lower in the respective groups, there is also a filled f^{14} electron set. As the metal is in a high oxidation state, the bonding in each compound is predominantly covalent. As examples, Table 9.1 shows oxo-anions of the 4th Period of Group 5, Group 6, and Group 7, together with the corresponding pseudo-isoelectronic oxo-anions of the 3rd Period and 4th Period of Group 15, Group 16, and Group 17.

Group 3 and Group 13

It was Rang in 1893 who seems to have been the first, on the basis of chemical similarity, to place boron and aluminum in Group 3 (see Figure 9.3) [4].

Such an assignment seems to have been forgotten until more recent times. Greenwood and Earnshaw [14] have discussed the way in which aluminum can be considered as belonging to Group 3 as much as to Group 13 (Figure 9.4), particularly in its physical properties. Habashi has suggested that there are so many similarities between aluminum and scandium that aluminum's place in the Periodic Table should actually be shifted to Group 3 [15].

Valence	I	II	III	IV
Series				
1	"	"	"	"
2	Li	Be	B	C
3	Na	Mg	Al	Si
4	K	Ca	Sc	Ti
5	Rb	Sr	Y	Zr
6	Cs	Ba	Di	"
7	"	Ms	"	Th

Figure 9.3 The first section of Rang's Periodic Table showing the location of boron and aluminum (from Ref. [4]).

Group 3	Group 13
	Al
Sc	Ga
Y	In
La	Tl

Figure 9.4 Members of Group 3 and Group 13.

Table 9.2 A comparison of standard reduction potentials for the Group 3 and 13 elements

Group 3		Group 13	
Element	E^θ (V)	Element	E^θ (V)
		Aluminum	−1.66
Scandium	−1.88	Gallium	−0.53
Yttrium	−2.37	Indium	−0.34
Lanthanum	−2.52	Thallium	+0.72

In terms of the electron configuration of the tripositive ions, one would indeed expect that Al^{3+} (electron configuration, [Ne]) would resemble Sc^{3+} (electron configuration, [Ar]) more than Ga^{3+} (electron configuration, $[Ar]3d^{10}$). Also of note, the standard reduction potential for aluminum fits better with those of the Group 3 elements than the Group 13 elements (Table 9.2) — as does its melting point.

Table 9.3   A comparison of aluminum, scandium, and gallium species under oxidizing conditions

	Acidic	Mid-to-Basic pH Range	Very Basic
Aluminum	$[Al(OH_2)_6]^{3+}(aq)$	$Al(OH)_3(s)$	$[Al(OH)_4]^-(aq)$
Scandium	$[Sc(OH_2)_6]^{3+}(aq)$	$Sc(OH)_3(s)$	$[Sc(OH)_4]^-(aq)$
Gallium	$[Ga(OH_2)_6]^{3+}(aq)$	$Ga(OH)_3(s)$	$[Ga(OH)_4]^-(aq)$

In terms of their comparative solution behavior, aluminum resembles both scandium(III) and gallium(III). For each ion, the free hydrated cation exists only in acidic solution. On addition of hydroxide ion to the respective cation, the hydroxides are produced as gelatinous precipitates. Each of the hydroxides redissolve in excess base to give an anionic hydroxo-complex, $[M(OH)_4]^-$. The similarities are summarized in Table 9.3 (data for this, and later tables, from Ref. [16]).

There does seem to be a triangular relationship between these three elements. However, aluminum does more closely resemble scandium rather than gallium in its chemistry. If hydrogen sulfide is bubbled through a solution of the respective cation, scandium ion gives a precipitate of scandium hydroxide, and aluminum ion gives a corresponding precipitate of aluminum hydroxide. By contrast, gallium ion gives a precipitate of gallium(III) sulfide. Also, scandium and aluminum both form carbides, while gallium does not.

Not previously identified, there are similarities in the chemistry of yttrium and indium. For example, their aqueous chemistry is dominated by the soluble 3+ cation in acid and by the insoluble hydroxide at neutral and basic pH (Table 9.4).

As a final note, the bottom member of Group 13, thallium, has very different chemistry to either yttrium or indium. The chemistry of thallium is more appropriately

Table 9.4 A comparison of yttrium and indium species under oxidizing conditions

	Acidic	Mid-pH Range	Basic
Yttrium	$[Y(OH_2)_6]^{3+}(aq)$	$Y(OH)_3(s)$	
Indium	$[In(OH_2)_6]^{3+}(aq)$	$In(OH)_3(s)$	

linked to that of silver through the "knight's move" relationship (see Chapter 10).

Group 4 and Group 14

This Group–pair (Figure 9.5) seems to be unique in that, although there are similarities between titanium(IV) and silicon(IV), there is a much greater resemblance of titanium(IV) with tin(IV), farther down Group 14.

In some ways, the chemistry of titanium(IV) resembles that of all the Group 14 elements in their +4 oxidation state. In particular, all five form tetrahedrally coordinated chlorides that are hydrolyzed to give the dioxide and hydrogen chloride.

Interestingly, even though they have significantly different molar masses, the chlorides of titanium(IV) and tin(IV) have remarkably similar melting and boiling points (Table 9.5). By contrast, both zirconium(IV) chloride and hafnium(IV) chloride are high-melting solids with a polymeric, six-coordinate structure.

As another example of the similarity of titanium(IV) and tin(IV), the most common form of crystal structure of titanium(IV) oxide is *rutile*, and tin(IV) oxide adopts the same structure. Also, titanium(IV) oxide and tin(IV) oxide

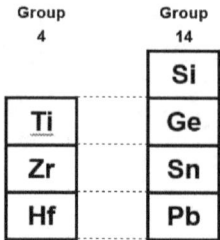

Figure 9.5 Members of Group 3 and Group 13.

Table 9.5 A comparison of melting points for the Group 4 and 14 chlorides

Group 4		Group 14	
Compound	M.Pt. (°C)	Compound	M.Pt. (°C)
–	–	$SiCl_4$	−70
$TiCl_4$	−24	$GeCl_4$	−52
$ZrCl_4$	+437	$SnCl_4$	−33
$HfCl_4$	+432	$PbCl_4$	−15

share the rare attribute of thermochromism by turning from white to yellow reversibly on heating.

Although the emphasis in this chapter is on links near the top of the respective groups, there are also relationships among the lower members. For example, the zirconates, $M_2Zr_2O_7$, and the stannates, $M_2Sn_2O_7$, adopt similar structures [17].

Group 5 and Group 15

In this Group–pair (Figure 9.6), the major resemblance seems to be between vanadium(V), phosphorus(V), and

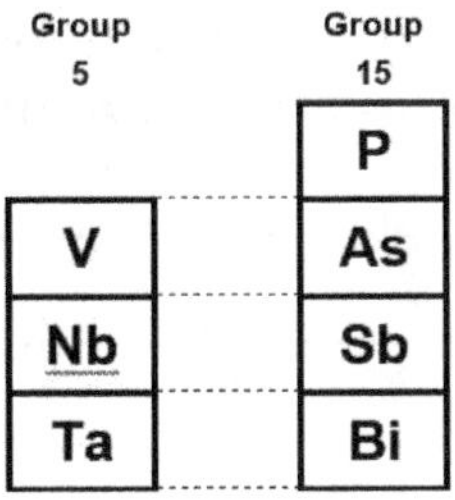

Figure 9.6 Members of Group 5 and Group 15.

Table 9.6 A comparison of aqueous vanadium, phosphorus, and arsenic species in dilute solution under oxidizing conditions

	Very Acidic	Acidic	Basic	Very Basic
Vanadium	VO_2^+	$H_2VO_4^-$	HVO_4^{2-}	VO_4^{3-}
Phosphorus	H_3PO_4	$H_2PO_4^-$	HPO_4^{2-}	PO_4^{3-}
Arsenic	H_3AsO_4	$H_2AsO_4^-$	$HAsO_4^{2-}$	AsO_4^{3-}

arsenic(V). There is also a similarity in aqueous species between niobium and antimony.

In terms of the simple oxo-anions, vanadium resembles both phosphorus and arsenic. Vanadate, phosphate, and arsenate are all strong bases with similar pKa values. The only significant difference is that at low pH, vanadium forms the vanadyl ion, not the undissociated acid as do phosphorus and arsenic (Table 9.6).

There are a significant number of parallel compounds between vanadium, phosphorus, and arsenic, as can be seen from Table 9.7. Of the two, vanadium more closely resembles phosphorus as there are several examples, two of which are listed in the following, for which there is no known arsenic analogue.

Table 9.7   Some parallel compounds and ions of vanadium(V) with phosphorus(V) and arsenic(V)

Group 7	Group 17
V_2O_5	P_4O_{10}, As_2O_5
VO_4^{3-}	PO_4^{3-}, AsO_4^{3-}
$V_4O_{12}^{4-}$	$P_4O_{12}^{4-}$
$VOCl_3$	$POCl_3$, $AsOCl_3$
VF_5	PF_5, AsF_5
VF_6^-	PF_6^-, AsF_6^-
VOF_4^-	POF_4^-(unstable)

Table 9.8   A comparison of aqueous antimony and niobium species under oxidizing conditions

	Acidic	Mid-pH Range	Basic
Niobium	$Nb_2O_5(s)$		Isopolyniobates(*aq*)
Antimony	$Sb_2O_5(s)$	Isopolyantimonates(*aq*)	

There is also an interesting parallel in aqueous species between antimony and niobium as is shown in Table 9.8. Though the aqueous antimonate ion is usually represented as [Sb(OH)$_6$]$^-$(*aq*) and the aqueous niobate ion as [NbO$_3$]$^-$ (*aq*), Greenwood and Earnshaw [18] have pointed out that for both of them, isopolymeric species predominate over most of the soluble range. By contrast, tantalum forms insoluble tantalum(V) oxide across the full pH range, while bismuth(III) dominates that element's aqueous chemistry.

The 4th Period Anomaly Revisited

In Chapter 7, the concept of the 4th Period anomaly was introduced. This anomaly was characterized by some aspects of the chemistry of the 4th Period member of a specific group differing from the pattern for the other group members [19]. Dasent listed some of the 4th Period examples [20]. In this context, he noted that while PCl_5, $NbCl_5$ (of Group 5), and $SbCl_5$ (of Group 15) are stable and well characterized, members of the 5th Period, both VCl_5 (Group 5) and $AsCl_5$ (Group 15) were elusive. They are now known, $AsCl_5$ decomposing above −50°C [21] and VCl_5 decomposing above −40°C [22], but certainly not "stable" species like the other matching chlorides.

Group 6 and Group 16

Just as vanadium(V) resembles phosphorus(V) and arsenic(V), so chromium(VI) resembles both sulfur(VI) and selenium(VI) (Figure 9.7).

Again there are parallels in the acid–base behavior of the oxo-anions, the only difference in this case being that chromic acid is a weaker acid than either sulfuric acid or selenic acid (Table 9.9).

There are also several formula similarities between chromium(VI) and both sulfur(VI) and selenium(VI). A few examples are given in Table 9.10.

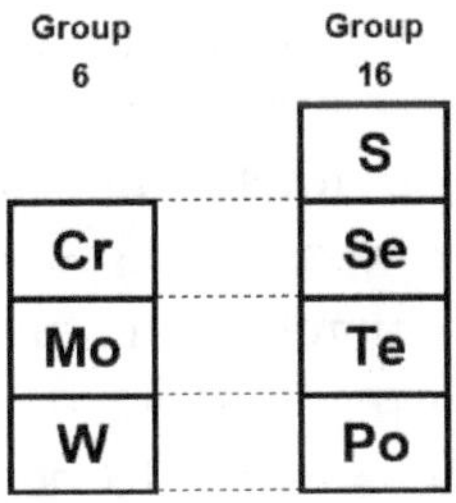

Group 6	Group 16
	S
Cr	Se
Mo	Te
W	Po

Figure 9.7 Members of Group 6 and Group 16.

Table 9.9 A comparison of aqueous chromium, sulfur, and selenium species under oxidizing conditions

	Very Acidic	Acidic	Basic	Very Basic
Chromium	H_2CrO_4	$HCrO_4^-$	CrO_4^{2-}	
Sulfur	HSO_4^-	SO_4^{2-}		
Selenium	$HSeO_4^-$	SeO_4^{2-}		

Table 9.10 Some parallel compounds and ions of chromium(VI) with sulfur(VI) and selenium(VI)

Group 7	Group 17
CrO_3	SO_3, SeO_3
CrO_4^{2-}	SO_4^{2-}, SeO_4^{2-}
$Cr_2O_7^{2-}$	$S_2O_7^{2-}, Se_2O_7^{2-}$
CrO_2Cl_2	SO_2Cl_2
CrF_6	SF_6, SeF_6
$CrOF_4$	$SOF_4, SeOF_4$
CrO_2F_2	SO_2F_2, SeO_2F_2

Group 7 and Group 17

Once again, there seems to be a triangular relationship, this time between manganese(VII), chlorine(VII), and bromine(VII) (Figure 9.8). There are also parallels in formulas between rhenium(VII) and iodine(VII).

The most obvious similarity between the three at the top of their Groups are the strongly oxidizing oxo-anions: permanganate, perchlorate, and perbromate. All three elements form corresponding trioxofluorides: MnO_3F, ClO_3F, and BrO_3F. There seems to be a slightly greater similarity between manganese(VII) and chlorine(VII) in that only those two form oxides in the +7 oxidation state: Cl_2O_7, and

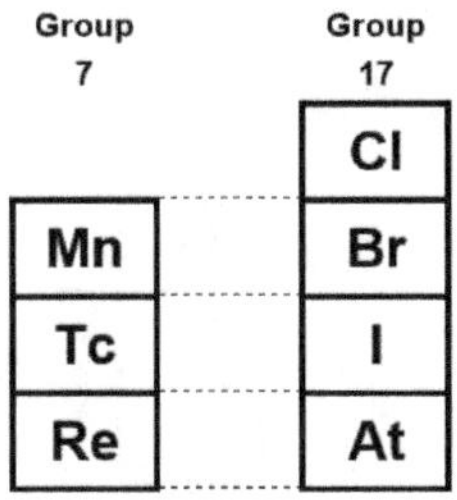

Figure 9.8 Members of Group 7 and Group 17.

Table 9.11 Some parallel compounds and ions of rhenium(VII) and iodine(VII)

Group 7	Group 17
Re_2O_7	I_2O_7
ReF_7	IF_7
$ReOF_5$	IOF_5
$ReO_2F_4^-$	$IO_2F_4^-$

Mn_2O_7, both of which are highly explosive liquids at room temperature.

Equally interesting in this group–pair are the similarities of rhenium with iodine.

Some of the parallel compounds are shown in Table 9.11. Of note, $ReOF_5$ has a melting point of 44°C while that of IOF_5 is 45°C.

Group 8 and Group 18

Somewhat surprisingly, there are two elements near the bottom of each group that share several similarities in chemical formulas: "noble metal" osmium(VIII) and "noble gas" xenon(VIII) (Figure 9.9).

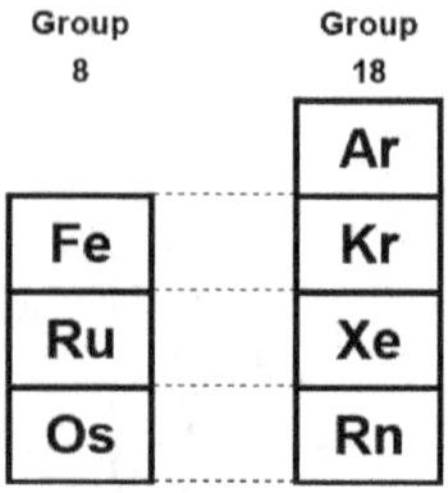

Figure 9.9 Members of Group 8 and Group 18.

Table 9.12 Some parallel compounds of osmium(VIII) and xenon(VIII)

Group 8	Group 18
OsO_4	XeO_4
OsO_3F_2	XeO_3F_2
OsO_2F_4	XeO_2F_4

Some parallels in formula are shown in Table 9.12. Similarities even extend to chemical behavior: osmium(VIII) oxide, OsO_4, is a yellow solid and strongly oxidizing; while xenon tetraoxide, XeO_4, is a pale yellow explosive compound.

Group 1 and Group 11

Group 1 and Group 11 are the most problematic of the group–pairs. In Chapter 8, it was argued that, in fact, there is no "Group 11" as the three elements: copper, silver, and gold, have so little in common. In the context of this chapter, it is the parallel in the +1 oxidation state, which must be considered. It is only for silver that this oxidation state dominates, thus it will be the focus of the comparison here.

There is one similarity: several of the sodium and silver(I) compounds are isostructural. These pairs include $NaNO_3$ and $AgNO_3$; Na_2SO_4 and Ag_2SO_4; and $Na_2S_2O_6 \cdot 2H_2O$ and $Ag_2S_2O_6 \cdot 2H_2O$.

This brief list provides meager evidence of linkage between these two Groups. Thompson alluded to the lack of any correlations [23]:

> Because of these differences in electronic structure comparison of silver with the alkali metals is fruitless even though each possesses a single s electron in the outermost shell. Probably the only similarity between the two is their diamagnetism and lack of colour.

Group 2 and Group 12

Based upon chemical similarities, in 1905, Werner designed a Periodic Table that showed beryllium and magnesium to belong to the zinc group (Figure 9.10) [5]. This was not mere speculation. Over 100 years later, Restrepo has shown on

							...	He
	Be	B	C	N	O	F	Ne	
	Mg	Al	Si	P	S	Cl	A	
Ni	Cu	Zn	Ga	Ge	As	Se	Br	Kr
Pd	Ag	Cd	Jn	Sn	Sb	Te	J	Xn
Pt	Au	Hg	Tl	Pb	Bi	...	...	...

Figure 9.10 Part of Werner's Periodic Table showing beryllium and magnesium as part of the zinc Group (from Ref. [5]).

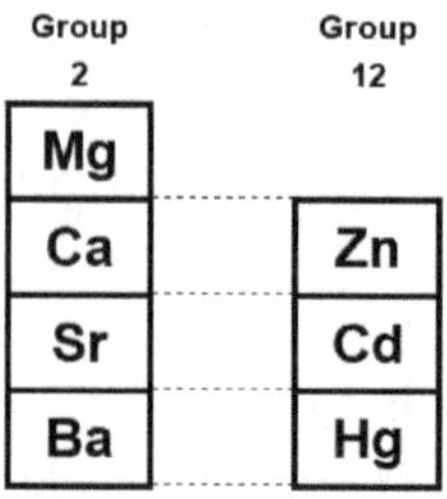

Figure 9.11   Members of Group 2 and Group 12.

chemical topological grounds, by means of a hypergraph, that the chemistry of beryllium and magnesium fits more closely with that of zinc, cadmium, and mercury, than the lower members of Group 2 [24].

There are certainly grounds to consider this assignment. All of these elements have +2 as the sole common oxidation state, except for mercury for which it is the higher oxidation state. Thus there are some resemblances for all of the Group 2 elements with zinc and cadmium (Figure 9.11). For example, all six of these elements form hygroscopic anhydrous metal chlorides.

The closer link seems to be between cadmium and calcium. Cadmium oxide has the NaCl structure, as do the Group 2 oxides. Of specific biochemical relevance, high levels of calcium ion inhibit the toxicity of cadmium(II) ion, suggesting that calcium and cadmium ions share the same cellular pathway [25]. Thus overall, cadmium and calcium seem to have the closer resemblance.

A Curious ($n + 5$) and ($n + 10$) Case

In this chapter, it has been shown how there are resemblances between scandium in Group 3 and aluminum in

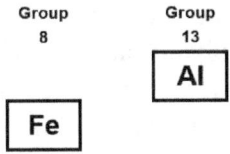

Figure 9.12   Iron of Group 8 and aluminum of Group 13.

Group 13. Curiously, the chemistry of aluminum also resembles that of the iron(III) ion (Figure 9.12). These similarities may be ascribed to the same 3+ charge and near-identical ion radii (and hence charge density). As a result of the high charge density, the $[M(OH_2)_6]^{3+}$ ions of both metals are very strongly acidic through hydrolysis.

There are some very specific similarities. For example, in the vapor phase, both ions form covalent chlorides of the form M_2Cl_6. These (anhydrous) chlorides can be used as Friedel–Crafts catalysts in organic chemistry, where they function by the formation of the $[MCl_4]^-$ ion [26]. In addition, another result of their high charge densities.

There are, however, some significant differences. For example, the amphoteric aluminum oxide reacts with hydroxide ion to give the soluble tetrahydroxoaluminate ion, $[Al(OH)_4]^-$, whereas the basic iron(III) oxide remains in the solid phase. It is this difference that enables aluminum oxide to be separated from iron(III) oxide in the commercial Beyer process, prior to the aluminum smelting step [27].

Commentary

Thanks to the work by Laing, the forgotten links between main group elements and transition metals have been rediscovered. Even though the eight-Group Table has long since been consigned to ancient history, these links in formula of compounds and polyatomic ions of elements of Group (n) with pseudo-isoelectronic species of elements of the corresponding Group ($n + 10$) provide chemists with a different perspective. Who would ever have imagined, for example, that there would be chemical similarities of osmium(VIII) and xenon(VIII)? What other unusual matching pairs are yet to be synthesized?

References

1. C. J. Giunta, "J. A. R. Newland's Classification of the Elements: Periodicity, but No System," *Bull. Hist. Chem.* **24**, 24–31 (1999).
2. E. R. Scerri, "A Philosophical Commentary of Giunta's Critique of Newlands' Classification of the Elements," *Bull. Hist. Chem.* **26**(2), 124–129 (2001).
3. G. N. Quam and M. B. Quam, "Types of Graphic Classifications of the Elements: Introduction and Short Tables: Introduction and Short Tables," *J. Chem. Educ.* **11**, 27–32 (1934).
4. P. J. F. Rang, "The Periodic Arrangement of the Elements," *Chem. News* 178 (14 April 1893).
5. G. N. Quam and M. B. Quam, "Types of Graphic Classifications of the Elements: Long Charts," *J. Chem. Educ.* **11**, 217–223 (1934).
6. W. C. Fernelius and W. H. Powell, "Confusion in the Periodic Table of Elements," *J. Chem. Educ.* **59**(6), 504–508 (1982).
7. E. Fluck, "New Notations in the Periodic Table," *Pure Appl. Chem.* **60**(3), 431–436 (1988).
8. R. T. Sanderson, "One More Periodic Table," *J. Chem. Educ.* **31**(9), 481 (1954).
9. M. Laing, "The Periodic Table: A New Arrangement," *J. Chem. Educ.* **66**, 746 (1989).
10. M. Laing, "The Periodic Table — Again," *Educ. Chem.* **26**, 177–178 (1989).
11. R. L. Rich, "Are Some Elements More Equal than Others?" *J. Chem. Educ.* **82**, 1761–1763 (1991).

12. D. M. P. Mingos, *Essential Trends in Inorganic Chemistry*, Oxford University Press, Oxford, 196–202 (1998).

13. G. Rayner-Canham, "Periodic Patterns: The Group (*n*) and Group (*n* + 10) Linkage," *Found. Chem.* **15**, 229–237 (2013).

14. N. N. Greenwood and A. Earnshaw, *Chemistry of the Elements*, 2nd ed., Butterworth-Heinemann, Oxford, 946–948 (1997).

15. F. Habashi, "Metals: Typical and Less Typical, Transition and Inner Transition," *Found. Chem.* **12**, 31–39 (2010).

16. G. K. Schweizer and L. L. Pesterfield, *Aqueous Chemistry of the Elements*, Oxford University Press, Oxford (2010).

17. R. A. Chapman, D. B. Meadowcroft, and A. J. Walkden, "Some Properties of Zirconates and Stannates with the Pyrochlore Structure," *J. Phys. D. Appl. Phys.* **3**(3), 307–319 (1970).

18. Ref. 14, Greenwood and Earnshaw, 577, 987.

19. R. T. Sanderson, "An Explanation of Chemical Variations within Periodic Major Groups," *J. Am. Chem. Soc.* **74**(19), 4792–4794 (1952).

20. W. E. Dasent, "Textbook Errors, XIV: Arsenic(V) Chloride," *J. Chem. Educ.* **34**(11), 535–536 (1957).

21. K. Seppelt, "Arsenic Pentachloride, $AsCl_5$," *Angew. Chem. Int. Ed.* **15**(6), 377–378.

22. F. Tamadon and K. Seppelt, "The Elusive Halides VCl_5, $MoCl_6$, and $ReCl_6$," *Angew. Chem. Int. Ed.* **15**(6), 377–378.

23. N. R. Thompson, "Silver," in *The Chemistry of Copper, Silver and Gold (Pergamon Texts in Inorganic Chemistry, volume 17)*, Pergamon Press, Oxford, 83 (1973).

24. G. Restrepo, "Challenges for the Periodic Systems of Elements: Chemical, Historical and Mathematical Perspectives," *Chem. Eur. J.* **25**, 15430–15440 (2019).

25. H. Michibata, S. Sahara, and M. K. Kojima, "Effects of Calcium and Magnesium Ions on the Toxicity of Cadmium to the Egg of the Teleost, *Oryzias latipes*," *Environ. Res.* **40**, 110–114 (1986).

26. W. H. Miles, C. F. Nutaitis, and C. A. Anderton, "Iron(III) Chloride as a Lewis Acid in the Friedel-Crafts Acylation Reaction," *J. Chem. Educ.* **73**(3), 272 (1996).

27. F. Habashi, "A Short History of Hydrometallurgy," *Hydrometallurgy* **79**(1–2), 15–22 (2005).

Chapter 10

Chemical "Knight's Move" Relationship

In 1999, Laing reported on a previously unrecognized relationship between elements in the lower right quadrant of the Periodic Table. He named these linkages the chemical "Knight's Move" Relationship. The name was chosen as the linkages were between an element and the element one Period below and two Groups to the right, the classic knight's move in the game of chess. In this chapter, we will look at some of these connections, focusing especially upon the double pairs.

The discovery of all the relationships covered in the previous chapters date back to the 19th and early 20th centuries. This chapter is devoted to a correlation that was not spotted until late in the 20th century.

The Group (n)/Period (m) and Group ($n + 2$)/ Period ($m + 1$) Linkages

It was in the pages of *Education in Chemistry* that Laing first reported the *knight's move* linkages in the Periodic Table [1]. The article described similarities that he had spotted between an element in the lower right quadrant of the Periodic Table and the element in the next lower Period and two Groups to the right. Thus, the linkages can be defined as:

> The **knight's move relationship** represents a pattern of similarities in the lower right quadrant of the Periodic Table between an element of Group (n) and Period (m) with an element of Group ($n + 2$) and Period ($m + 1$).

Laing identified the sets of elements having atomic numbers: 29 to 31, 47 to 51, and 79 to 84, as marking the boundaries of the knight's move (see Figure 10.1). The three pairs upon which Laing focused were: zinc and tin; silver and thallium; and mercury. In addition, using the knight's

Group 11	Group 12	Group 13	Group 14	Group 15	Group 16
Cu	Zn	Ga			
Ag	Cd	In	Sn	Sb	
Au	Hg	Tl	Pb	Bi	Po
			Fl		

Figure 10.1 The elements with potential to form knight's move relationship pairs (plus flerovium, element 114) according to Laing (from Ref. [1]).

Figure 10.2 The White Knight in *Alice Through the Looking Glass* (from Ref. [4]).

move, he made predictions about chemistry of element 114 (flerovium).

The choice of the chess analogy was particularly interesting in that there was already a link between chess and chemistry. The link involved Oxford University chemist Augustus Vernon Harcourt [2]. Vernon Harcourt, a very affable, gentle, and forgetful chemist, is widely believed to have been the model or part of, for the White Knight in the story of *Alice Through the Looking-Glass* [3] (Figure 10.2).

Laing's Knight's Move (K-M) Claims

To support his claim of the existence of the knight's move (K-M) relationship, Laing provided evidence from widely different aspects of element chemistry. For example, he noted that for the zinc–tin K-M pair, their respective compounds are nonpoisonous, while for the cadmium–lead K-M pair, their respective compounds are extremely poisonous.

Knight's Move Links among the Elements

For the zinc–tin K-M pair, Laing noted that both these elements plated steel. Also, zinc and tin formed alloys with copper: brass (zinc and copper) and bronze (tin and copper). There is even a ternary alloy of 96% copper, 2% zinc, and 2% tin. Then for the tin–polonium K-M pair, Laing pointed out that the elements themselves had very similar melting points: tin at 232°C, and polonium at 254°C.

Knight's Move Links among Compound Melting Points

Laing compared melting and boiling points among pairs of K-M related compounds. A melting point pair from each K-M set that he quoted is given in Table 10.1.

Table 10.1 Comparisons of melting points for some K-M pairs (from Ref. [1])

Silver(I)–Thallium(I)	Cadmium(II)–Lead(II)	Zinc–Tin(II)	Gallium(III)–Antimony(III)
AgCl 445°C	CdI_2 385°C	$ZnCl_2$ 275°C	$GaCl_3$ 77°C
TlCl 429°C	PbI_2 412°C	$SnCl_2$ 247°C	$SbCl_3$ 73°C

Knight's Move Chemical Relationships

In chemical similarities, Laing pointed out that silver chloride and thallium(I) chloride are both water-insoluble compounds. However, they differ in that silver chloride reacts with ammonia to give the soluble linear $[Ag(NH_3)_2]^+$ ion while thallium chloride does not react with ammonia. Also, Laing noted that the crystal structures of zinc oxide and tin(II) oxide are similar.

Knight's Move Prediction of Properties for Element 114

Laing used the K-M concept to predict some properties of the then-undiscovered element 114 (now called flerovium). He made his predictions of flerovium chemistry by comparison with mercury, the K-M match. Laing predicted flerovium would be a metal with very low melting point and a density of about 16 g·cm^{-3}. By comparison with mercury, he stated that flerovium should form a chloride, $FlCl_2$, and an oxide, FlO (which would be thermally unstable).

Some chemistry of flerovium has since been reported. Eichler *et al.* state that [5]:

> *Identification of three atoms of element 114 in thermochromatography experiments . . . indicates that this element is at least as volatile as simultaneously investigated elements Hg, At, and element 112. This behaviour is rather unexpected for a typical metal of group 14.*

Laing's prediction of a similarity to mercury seems to have been confirmed.

Laing's Knight's Move Legacy

Laing's many contributions to the discussions on the Periodic Table are sprinkled throughout this book. On the basis of that one article, the knight's move has become accepted as a genuine periodic relationship, having been included in such resources as: a conference proceedings on the Periodic Table [6]; and a definitive work on the history and structure of the Periodic Table [7]. And now, in this work, a whole K-M chapter. It will be through the K-M that Laing's name will live on [8].

Reevaluation of the Knight's Move Relationship

An overriding and oft-forgotten point about the chemical elements is that each element is unique. It is this individuality that makes inorganic chemistry such an interesting but, at the same time, gargantuan field of study. Thus, in looking for relationships, one should not expect total congruence among the elemental behaviors; on the other hand, one should be hoping to find consistent patterns that are more than simple probability. For this reason, the knight's move concept needs to be tested by looking systematically and comprehensively at one or more pairs of elements.

The Significance of Oxidation Number

In the opinion of this Author, it is the compounds of the same oxidation states that provide the knight's move with its main validity. In fact, this matching of oxidation states — generally the lower one — seems to be the key feature of the linkages (Figure 10.3).

The K-M Silver(I) — Thallium(I) Similarities

The most intriguing example of the K-M relationship is that of silver(I) and thallium(I). Some of the similarities are as follows:

Group 11	Group 12	Group 13	Group 14	Group 15	Group 16
Cu (+1), +2	**Zn** +2	**Ga** +3			
Ag +1, (+2)	**Cd** +2	**In** (+1), +3	**Sn** +2, (+4)	**Sb** +3, +5	
Au +1, +3	**Hg** (+1), +2	**Tl** +1, +3	**Pb** +2, (+4)	**Bi** +3, (+5)	**Po** +2, +4

Figure 10.3 The common oxidation states of the "Knight's Move" elements with the less common oxidation state indicated in parentheses.

- For silver and thallium, unique in their respective Groups, the +1 state is stable and preferred in aqueous solution.
- Silver(I) and thallium(I) halides are whitish except for the iodides that are yellow.
- Silver(I) fluoride and thallium(I) fluoride are water soluble and all other silver and thallium(I) halides are insoluble.
- Unique among chromates, insoluble silver(I) chromate, Ag_2CrO_4, and thallium(I) chromate, Tl_2CrO_4 are both brick red in color.
- In the mineral, *crookesite*, $Cu_7(Ag,Tl)Se_4$, silver(I) and thallium(I) occupy the same lattice sites.
- Thallium(I) tetrafluoroborate, $TlBF_4$, has been proposed as a substitute reagent for silver(I) perchlorate, $AgClO_4$ [9].

A Thallium Detour

Earlier, the knight's move linkage of thallium(I) with silver(I) was explored. In Chapter 9, the limited (n) and (n + 10) relationship of silver with the alkali metals was briefly mentioned. Combining these two links, a most curious connection is that of thallium(I) with the heavier alkali metals. For example, thallium, like potassium, forms a hydroxide, TlOH, which is very water soluble to produce a very basic solution. Thallium(I) is also one of the cations that fits the large monopositive ion site in an alum as a substitute for an alkali metal ion [10].

Thallium(I) ion is highly poisonous. It enters the body through the potassium ion uptake pathways. Once absorbed, it is the attraction to sulfur ligands that provide thallium(I) with its toxicity (and difference from the alkali metal ions). In this way, the thallium(I) ion disrupts many cellular processes by interfering with the function of proteins that incorporate the sulfur-containing amino acid, cysteine [11].

Knight's Move Relationships among "Double Pairs"

In addition to simple pairs, there are also "double pairs" of K-M related elements. These are copper–indium/indium–bismuth and zinc–tin/tin–polonium, in which each central element has two other elements linked by potential Knight's Move relationships (Figure 10.4).

To summarize Laing's claims, he proposed that the following specific features indicate the existence of a knight's move pattern:

- Similarities of metal melting points.
- Patterns in compound formulas and structures.
- Parallels in melting and boiling points of compounds of the same formulation.

Group 11	Group 12	Group 13	Group 14	Group 15	Group 16
Cu	Zn				
		In	Sn		
				Bi	Po

Figure 10.4 The two sets of "double pair" K-M related elements, one double pair is in italic.

Here the three aspects will be examined in the context of the two "double pairs."

Are Metal Melting Points Irrelevant to K-M?

Laing noted in his paper [1] the similarity in melting points between tin (232°C) and polonium (254°C). Looking at the melting and boiling points of the first double pair of zinc–tin–polonium (see Table 10.2), there does seem to be a similarity of melting points (tin–polonium) and boiling points (zinc–polonium), though there is no systematic pattern involving all three metals.

The matching table for the copper–indium–bismuth double pair indicates that low melting points are characteristic of all the lower p-block metallic elements (see Table 10.3). In fact, the closest match in melting and boiling points does not come from knight's move pairs. Instead, by comparing Tables 10.2 and 10.3, the major similarity for the main group elements is controlled by Period. Thus, tin and indium have similar melting and boiling points; as do

Table 10.2 The phase change temperatures for the Zn–Sn–Po double pair

Element	Zinc	Tin	Polonium
Melting point (°C)	420	232	254
Boiling point (°C)	907	2623	962

Table 10.3 The phase change temperatures for the Cu–In–Bi double pair

Element	Copper	Indium	Bismuth
Melting point (°C)	1083	157	271
Boiling point (°C)	2570	2073	981

bismuth and polonium. Therefore, similarities in metal melting points and boiling points do not seem to be a defining K-M feature [12].

Copper(I)–Indium(I) and Indium(III)–Bismuth(III) Double T-M Links

It is always the lower oxidation state of the 5th Period element that matches an oxidation state of the 4th Period element. Then it is the higher oxidation state of the 5th Period element that matches the lower oxidation state of the 6th Period element. For the double T-M links here, the pairs are compounds of copper(I) and indium(I); and then corresponding compounds of indium(III) and bismuth(III). Four main sources of information have been used [13–16].

The usual aqueous oxidation state for copper is +2 while that for indium is +3. For both copper and indium, the +1 oxidation state is a comparative rarity. It is found mostly in insoluble solid-state species, though in itself, the existence of this matching oxidation state is notable in the context of the knight's move. Table 10.4 shows some of the simple copper(I) and corresponding indium(I) compounds known. Also of note, fluorides are unknown for both copper(I) and indium(I).

The comparative chemistry of indium(III) and bismuth(III) is more extensive. The parallels between indium(III) and bismuth(III) are particularly strong as +3 is the more common oxidation state for both elements. Bismuth

Table 10.4 Some corresponding copper(I) and indium(I) compounds

Copper(I)	$CuCl$	$CuBr$	CuI	Cu_2O	Cu_2S
Indium(I)	$InCl$	$InBr$	InI	In_2O	In_2S

exemplifies the phenomenon that the elements of the later 6th Period tend to favor the lower oxidation states over the highest. Some of the similarities are as follows:

- All tripositive halides and chalcogenides are known for both elements.
- Indium(III) and bismuth(III) form corresponding tetra- and hexa-coordinate halo-complex ions: MX_4^- and MX_6^{3-}.
- Indium(III) and bismuth(III) form isostructural alums: $M^IM^{III}(SO_4)_2 \cdot 12H_2O$, where M^I is a large monopositive ion and M^{III} is indium(III) or bismuth(III).
- Indium(III) and bismuth(III) form stable oxo-halides of matching formula, such as InOCl and BiOCl.

Zinc–Tin(II) and Tin(IV)–Polonium(IV) Double T-M Links

Though here we focus on knight's move resemblances, it is important to note that an element in this region can also possess similarities to elements elsewhere in the Periodic Table. Zinc may hold the record in this context. In Chapter 9, the similarity of zinc (Group 12) to magnesium (Group 2) by the (n) and ($n + 10$) relationship is discussed, a linkage also reported by Laing [17]; while Massey has pointed out similarities of zinc with beryllium (Group 2) in compound formulas [18]. Massey also found similarities of zinc with gallium in chemical behavior (though not formula) [19].

The comparative chemistry of the zinc–tin(II) certainly provides several similarities.

- Zinc and tin(II) exhibit amphoteric behavior, their hydroxides dissolving in excess hydroxide ion to form zincates and stannates, respectively.
- Aqueous solutions of their divalent chlorides hydrolyze to give insoluble Zn(OH)Cl and Sn(OH)Cl.

- In the presence of high chloride ion concentrations, the chlorides give parallel chloro-complex ions: $ZnCl_3^-$ and $SnCl_3^-$; and $ZnCl_4^{2-}$ and $SnCl_4^{2-}$.
- Zinc and tin(II) form dialkyls of the form R_2Zn and R_2Sn (though the zinc series tend to be monomeric while the tin(II) compounds are polymeric).

Though some compounds of polonium(VI) are known, polonium, like bismuth, prefers lower oxidation states. There are a wide range of compounds of polonium(IV) together with some compounds of polonium(II). It was noted by Brasted over 45 years ago [20] that polonium bore little resemblance in its chemistry to tellurium and instead that polonium(II) behaved more like lead(II) of Group 14. Curiously, in one respect, polonium(II) has a resemblance to zinc(II): that is, polonium forms volatile dimethylpolonium(II), $(CH_3)_2Po$ [21] analogous to $(CH_3)_2Zn$.

Despite Brasted's claim of a link between polonium(II) and lead(II), there seem to be more similarities between polonium(IV) and tin(IV), than polonium(IV) and lead(IV). Following are examples of some matching formulas.

- There are matching chlorides and hexachloro-ions, $SnCl_4$ and $PoCl_4$, and $[SnCl_6]^{2-}$ and $[PoCl_6]^{2-}$.
- The only solid stable nitrates of both metals correspond: $Sn(NO_3)_4$ and $Po(NO_3)_4$.
- There are matching oxides in the +4 oxidation state: SnO_2 and PoO_2.

Melting Points of Some Copper(I)–Indium(I) and Indium(III)–Bismuth(III) Halides

Laing [1] noted close-matching melting points among the following halide pairs: $AgCl/TlCl$; $AgBr/TlBr$; CdI_2/PbI_2;

Table 10.5 Melting points of corresponding copper(I) and indium(I) halides

	MCl	MBr	MI
Copper(I)	422	504	606
Indium(I)	211	285	364

Table 10.6 Melting points of corresponding indium(III) and bismuth(III) halides

	MF_3	MCl_3	MBr_3	MI_3
Indium(III)	1170	583	435	207
Bismuth(III)	649	234	219	409

$ZnCl_2/SnCl_2$; and $GaCl_3/SbCl_3$. The question arises whether such patterns are pervasive, or just found for a few selected cases. Here, as an example, the melting point series are provided for the halides of two double pairs, copper(I)–indium(I) (Table 10.5) and indium(III)–bismuth(III) (Table 10.6).

As can be seen from the data earlier, there are no clear patterns among these K-M pairs. Nor could any consistent pattern be found for any other K-M pairs.

The Knight's Move Relationship and the "Inert Pair" Effect

Having established that there is indeed a K-M relationship, the question needs to be asked as to the reason for it. Laing attempted to answer the question [1]:

> Is the "knight's move" merely a special case of the "inert pair effect" applied to metals with a d^{10} electron configuration?

The Inert Pair Effect

First, a digression onto the definition of the "inert pair" effect. This phenomenon was first described by Sidgwick in 1927 [22]. The observation was concisely explained by Orgel [23]:

> *Many B subgroup [main-group] metals exhibit a stable valency two smaller than the group valency. This tendency is most pronounced for thallium, lead, and bismuth and is also important for many lighter elements such as tin and antimony.*

The contention then, is that, in the cases of ionic bonding, the ns^2 electrons are significantly more tightly bound than the np^x electrons. Or to reverse the statement, the np^x electrons are more easily removed. For example, the noble gas core electron configuration of tin is: $[Kr]5s^2 4d^{10} 5p^2$; the configuration of the more common ion, Sn^{2+}, would be $[Kr]5s^2 4d^{10}$; and that of the less common ion, Sn^{4+}, would be $[Kr]4d^{10}$. The prevalence of the tin(II) ion would therefore be attributable to the inert pair effect.

However, with many of these compounds, the bonding is believed to be more covalent than ionic. Drago developed an explanation of the inert pair effect in terms of the low bond enthalpies of the heavy p-block metals [24]. Laing, himself, recognized this problem [1]:

> *There is more behind the knight's move than meets the eye. We are dealing here with an extremely complex phenomenon, not easy to explain. ... Nevertheless, application of the idea of the knight's move among metals with d^{10} configurations on the bottom right hand side of the Periodic Table leads to many correct predictions that would not be made by applying the usually accepted trends in the Periodic Table.*

"Inert Pair" as a Relativistic Effect

It is relativistic effects, first mentioned in Chapter 2, that provide the most logical explanations for most of the inert pair phenomenon [25–27]. When electron relativistic effects

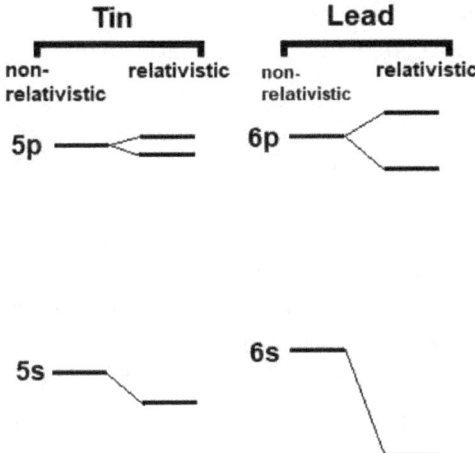

Figure 10.5 Comparative nonrelativistic and relativistic energies for the s- and p-orbitals of tin and lead (adapted from Ref. [26]).

are considered, the energy of the electrons in the s-orbital drops significantly, that is, the electrons are more tightly bound to the nucleus. This pattern is shown in Figure 10.5 for tin and lead.

As can be seen, there is a small (yet significant) decrease in the energy of the 5s orbital for tin while there is a dramatic decrease in the energy of the 6s orbital for lead, that is, the $6s^2$ electron pair is exceptionally strongly bound to the nucleus. Thus, it would seem that there is indeed a satisfactory explanation for most aspects of the knight's move relationship.

Commentary

The indication of a periodic pattern is the consistent applicability of a phenomenon to a subset of the Periodic Table of Elements. On this basis, the justification of the knight's move relationship should be made primarily on the basis of similarities in formulas and chemistry of compounds of knight's move pairs of elements in the lower right quadrant of the Periodic Table.

Though there are a few specific resemblances in melting and boiling points among pairs of "Knight's Move" compounds, they are not widespread and consistent enough to be regarded as evidence of a systematic pattern. Thus, it is the chemical, rather than the physical, properties that should be emphasized as evidence for this relationship.

References

1. M. Laing, "The Knight's Move in the Periodic Table," *Educ. Chem.* **36**, 160–161 (1999).
2. J. Shorter, "Vernon Harcourt: A Founder of Chemical Kinetics and a Friend of 'Lewis Carroll'," *J. Chem. Educ.* **57**, 411–416 (1980).
3. M. C. King, "The Chemist in Allegory: Augustus Vernon Harcourt and the White Knight," *J. Chem. Educ.* **60**, 177–180 (1983).
4. L. Carroll, *More Annotated Alice: Alice's Adventures in Wonderland and Through the Looking Glass and What Alice Found There*, with notes by Martin Gardner; Random House, New York, 277 (1990).
5. R. Eichler *et al.*, "Indication for a Volatile Element 114," *Radiochim. Acta* **98**, 133–139 (2010).
6. D. H. Rouvray and R. B. King (eds.), *The Periodic Table: Into the 21st Century*, Research Studies Press Ltd., Hertfordshire, England, 135–136, 177–179 (2004).
7. E. R. Scerri, *The Periodic Table: Its Story and Its Significance*, Oxford University Press, New York, NY, 272–275 (2007).

8. G. B. Kauffman and L. M. Kauffman, "Michael J. Laing (1937–2012), Obituary of a Lovable, Inspirational Inorganic Chemist, Chemical Educator, and Science Popularizer," *Chem. Educ.* **20**, 183–193 (2015).

9. F. J. Arnáiz, "The Preparation of TlBF$_4$," *J. Chem. Educ.* **74**(11), 1332–1333 (1997).

10. J. L. Lambert and M. W. Lambert, "The Alums: Interchangeable Elements in a Versatile Crystal Structure," *J. Chem. Educ.* **47**(6), 465 (1970).

11. K. T. Douglas, M. A. Bunni, and S. R. Baindur, "Thallium in Biochemistry," *Int. J. Biochem.* **22**(5), 429–438 (1990).

12. G. Rayner-Canham and M. Oldford, "The Chemical 'Knight's Move Relationship: What Is Its Significance?" *Found. Chem.* **9**, 119–125 (2007).

13. J. C. Bailar *et al.*, (eds.), *Comprehensive Inorganic Chemistry*, Pergamon Press, Oxford (1973).

14. F. A. Cotton, G. Wilkinson, C. A. Murillo, and M. Bochmann, *Advanced Inorganic Chemistry*, 6th ed., Wiley-Interscience, New York, NY (1999).

15. N. N. Greenwood and A. Earnshaw, *Chemistry of the Elements*, 2nd ed., Butterworth-Heinemann, Oxford (1997).

16. A. G. Massey, *Main Group Chemistry*, 2nd ed., John Wiley, Chichester (2000).

17. M. Laing. "The Periodic Table — A New Arrangement," *J. Chem. Educ.* **66**, 746 (1989).

18. Ref. 16, A. G. Massey, p. 174.

19. Ref. 16, A. G. Massey, p. 208.

20. R. C. Brasted, *Comprehensive Inorganic Chemistry*, vol. 8, Van Nostrand, New York, NY, 250 (1961).

21. A. S. Bahrou *et al.*, "Volatile Dimethyl Polonium Produced by Aerobic Marine Microorganisms," *Environ. Sci. Technol.* **46**(20), 11402–11407 (2012).

22. N. V. Sidgwick, *The Electronic Theory of Valency*. Clarendon Press, Oxford, 178–181 (1927).

23. L. E. Orgel, "The Stereochemistry of B Subgroup Metals. Part II. The Inert Pair," *J. Chem. Soc.* 3815–3819 (1959).

24. R. S. Drago, "Thermodynamic Evaluation of the Inert Pair Effect," *J. Phys. Chem.* **62**(3), 353–357 (1958).

25. K. S. Pitzer, "Relativistic Effects on Chemical Properties," *Acc. Chem. Res.* **12**, 271–276 (1977).

26. P. Pyykkö, "Relativistic Effects in Structural Chemistry," *Chem. Rev.* **88**(3), 563–594 (1988).

27. J. S. Theyer, "Relativistic Effects and the Chemistry of the Heaviest Main-Group Elements," *J. Chem. Educ.* **82**, 1721–1727 (2005).

Chapter 11

Isodiagonality

Diagonal relationships in the Periodic Table were recognized by both Mendeléev and Newlands. More appropriately called isodiagonal relationships, the same three examples of lithium with magnesium; beryllium with aluminum; and boron with silicon; are commonly quoted. Here these three pairs of elements are discussed in detail, together with evidence of isodiagonal linkages elsewhere in the Periodic Table.

Though the vertical groups and horizontal periods are emphasized as being the key relationships in the Periodic Table (see Chapters 7 and 8), chemistry historian Ihde has noted that both Mendeléev and Newlands had reported the diagonal similarities of lithium with magnesium; beryllium with aluminum; and boron with silicon; as early as 1860 [1].

Isodiagonality

It was in 1937 that French explored the diagonal relationship in detail. In fact, he considered diagonality to extend further down the Periodic Table [2]:

In chemical behavior, bismuth bears a much greater resemblance to silicon and boron than to nitrogen.

This diagram, from French's article, was probably the first to illustrate the diagonal linkages (Figure 11.1).

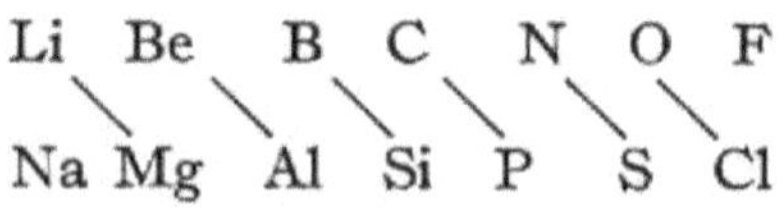

Li Be   B   C   N   O   F
   \   \   \   \   \   \
  Na Mg Al Si   P   S   Cl

Figure 11.1 French's diagram of the isodiagonal linkages (from Ref. [2]).

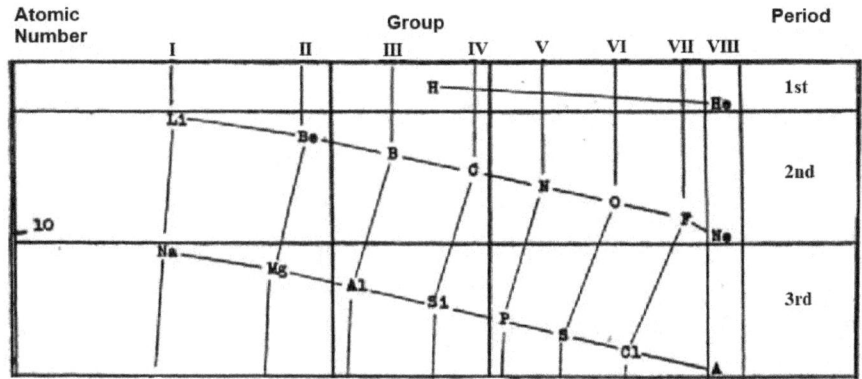

Figure 11.2 The first three Periods of French's slanting periodic table to illustrate group and diagonal linkages (adapted from Ref. [2] — note "A" was the former symbol for argon).

French was concerned that diagonality not be overemphasized, adding:

> ... *silicon for instance, resembles neither carbon nor boron in chemical properties but might be said to lie between the two.*

Thus, French suggested that the best mode of display of the "cross relationships" might be that shown in his "warped" Periodic Table in the following (Figure 11.2).

More recently, Rich proposed the term *isodiagonal* for species related on an upper-left to lower-right diagonal [3]. He subsequently authored a version of the Periodic Table to specifically emphasize isodiagonalities (Figure 11.3) [4].

Here the following definition of isodiagonal will be used:

> *An **isodiagonal relationship** is identified by similarity in chemical properties between an element and that to the lower right of it in the periodic table.*

Isodiagonality is, in some ways, a general attribute of the properties of the chemical elements. For example,

			1	2		3	4												
			H	He		Li	Be												
5	6	7	8	9	10	11	12	13											
B	C	N	O	F	Ne	Na	Mg	Al											
	14	15	16	17	18	19	20	21	22	23	24	25	26	27	28	29	30	31	32
	Si	P	S	Cl	Ar	K	Ca	Sc	Ti	V	Cr	Mn	Fe	Co	Ni	Cu	Zn	Ga	Ge
	33	34	35	36	37	38	39	40	41	42	43	44	45	46	47	48	49	50	51
	As	Se	Br	Kr	Rb	Sr	Y	Zr	Nb	Mo	Tc	Ru	Rh	Pd	Ag	Cd	In	Sn	Sb
	52	53	54	55	56	57–71	72	73	74	75	76	77	78	79	80	81	82	83	84
	Te	I	Xe	Cs	Ba	La–Lu	Hf	Ta	W	Re	Os	Ir	Pt	Au	Hg	Tl	Pb	Bi	Po
	85	86	87	88	89–103	104	105	106	107	108	109	110	111	112	113	114	115	116	117
	At	Rn	Fr	Ra	Ac–Lr	Rf	Db	Sg	Bh	Hs	Mt	Ds	Uuu	Uub	Uut	Uuq	Uup	Uuh	Uus

Figure 11.3 Rich's version of the periodic table to emphasize isodiagonal relationships (from Ref. [4]).

Edwards and Sienko commented that the metal–nonmetal divide forms an "almost diagonal demarcation" [5]. Similarly, the elements often considered to be semimetals fall on a roughly diagonal boundary between the metals and nonmetals, see Chapter 5 [6].

A related phenomenon, the change in bonding type across Periods, similarly lies upon a diagonal [7]. The pattern is usually for a change from ionic (to the left) to small-molecule covalent (to the right) with a species that can be assigned as possessing network covalent bonding at the transition point. As was shown in Chapter 7, for 2nd Period and 3rd Period oxides, this intermediate bond type occurs with B_2O_3 and SiO_2. For fluorides, the transition is displaced left by one group so that it occurs with BeF_2 and AlF_3, and similarly for hydrides with $(BeH_2)_x$ and $(AlH_3)_x$.

Explanations for Isodiagonality

Cartledge, in 1928, was the first to suggest a possible explanation for isodiagonality [8]. He proposed that the phenomenon could be explained in terms of ionic potential, Z/r, what is now more commonly known as charge–radius ratio. The ionic potential was recalculated by Hanusa using Shannon–Prewitt ionic radii and these values correlated well with isodiagonal links for some pairs, but not others [9]. As an example, the ionic potential for Be^{2+} of 74 nm^{-1} is very close to the value of 77 nm^{-1} for Al^{3+}. However, there is no match in the values for the pair of Li^+ (17 nm^{-1}) and Mg^{2+} (35 nm^{-1}). The same ratio, but called polarizing power, was qualitatively used to explain isodiagonality by Puddephatt and Monaghan [10].

Lee provided a variety of explanations [11]. He first proposed polarizing power, then suggested radius similarities for the Li^+–Mg^{2+} link; and charge per unit area to explain the Be^{2+}–Al^{3+} link; but commented that electronegativity similarities was another possible explanation. Finally, in the context of the Be^{2+}–Al^{3+} link, Lee stated:

> Just as was the case with lithium and magnesium, the similarity in atomic and ionic sizes is the main factor underlying this relationship.

King also favored electronegativity as an explanation [12]. Housecroft and Sharpe, by contrast, proposed that isodiagonality could be explained in terms of similarities in ionic radius [13]. However, this contradicted Hanusa's conclusions, as ion radius and ionic potential are reciprocal relationships.

Rayner-Canham and Overton found that *charge density* is a useful parameter for predicting ionic versus covalent behavior in simple binary compounds and that it could also account for the diagonal Li^+–Mg^{2+} and Be^{2+}–Al^{3+} links [14]. This term, charge density, dates back to at least the 1960s [15]. It is defined as:

> The **charge density** of a real or hypothetical ion is defined as the ion charge divided by the ion volume.

In order to obtain numbers in meaningful units and, at the same time, avoid the need for enormous exponents, Rayner-Canham and Overton utilized the electron charge in Coulombs and the ionic radius in millimeters. Thus, for each real or theoretical ion, the integer ion charge was multiplied by the electron charge and divided by $4/3\pi$ times the ion volume to give values in $C \cdot mm^{-3}$.

Unfortunately, some sources confuse charge density with charge–radius ratio. For example, Rogers stated that charge density was defined as "charge on a metal cation over its ionic radius" [16]. This parameter is, in fact, correctly named charge–radius ratio.

Despite this confusion, Rogers provides one of the more comprehensive discussions of isodiagonality. He correlated the parameters of ionic radius, charge–radius ratio, and electronegativity for the Li–Mg, Be–Al, and B–Si pairs:

> There appear to be three principal factors why these pairs — take beryllium and aluminum as a representative example — have so much chemistry in common. One factor is ionic size; the others are charge density (or charge-radius-ratio, Z/r) and electronegativity. ... The two metal ions, then, will similarly polarize the X atom in an M–X bond and give rise to a similar additional covalent character on that basis.

He added the caveats:

> First, keep in mind that group relationships (for example, between beryllium and magnesium) are still the dominant factor. ... Second, the ions ... particularly the highly charged B^{3+}, C^{4+}, and Si^{4+} really do not exist as such. ... Nevertheless, even with these warnings, the diagonal relationship remains a good organizing principle.

Table 11.1 shows some of the parameters for these eight ions. The ionic radii in pm (Shannon–Prewitt), both for 4- and 6-coordination are from Ref. [17]; charge–radius ratio values are calculated from the Shannon–Prewitt ion radius values for 6-coordination (in nm^{-1}); charge density values are from Ref. [14] (in $C \cdot mm^{-3}$); and Allred–Rochow electronegativity values are from Ref. [18].

It is difficult to attribute any one parameter as a ubiquitous explanation for all isodiagonal resemblances. This is

Table 11.1   Parameter values for the early 2nd Period and 3rd Period elements

	Group 1	Group 2	Group 13	Group 14
	Lithium	Beryllium	Boron	Carbon
Charge of ion	+1	+2	+3	+4
Ionic radius (4)	59	27	11	15
Ionic radius (6)	76	45	27	16
Charge–radius ratio	13	44	111	250
Charge density	98	1108	1663	6265
Electronegativity	0.97	1.47	2.01	2.50
	Sodium	Magnesium	Aluminum	Silicon
Charge of ion	+1	+2	+3	+4
Ionic radius (4)	99	57	39	26
Ionic radius (6)	102	72	54	40
Charge–radius ratio	11	28	56	100
Charge density	24	120	770	—
Electronegativity	1.01	1.23	1.47	1.74

not surprising, considering the Li–Mg pair are predominantly ionic in behavior while B–Si are totally covalent in their properties. In the following sections, individual pairs of elements will be compared and contrasted in terms of possible isodiagonal relationships.

Isodiagonality of Lithium and Magnesium

Though lithium and magnesium are often taken as the prototypical isodiagonal pair, it more highlights the 2nd Period Anomaly: that the 2nd Period elements are uniquely different to the lower members of their group (see Chapter 7). As we see in the following, there are indeed specific similarities between lithium and magnesium, though on the basis of free energies of formation of compounds, Hanusa found a closer resemblance of lithium with calcium [9]. On the other hand, Greenwood and Earnshaw contended that magnesium is atypical of Group 2 (though beryllium is even more so) and that lithium does match well and uniquely with magnesium [19].

General Resemblance of Lithium to Group 2 Elements

First, there are resemblances between lithium and the Group 2 elements as a whole:

- Lithium does not form an isolatable hydrogen carbonate whereas solid hydrogen carbonate salts can be obtained for the other Group 1 metals. Solid hydrogen carbonates cannot be isolated for the Group 2 metals.
- Lithium salts tend to be hydrated (often as a trihydrate) whereas the salts of the other Group 1 elements tend to be anhydrous. Many Group 2 metal salts are hydrated.
- Three lithium salts — carbonate, phosphate, and fluoride — have very low solubility unlike the salts of the other Group 1 metals. These anions form insoluble salts with the Group 2 metals.
- Lithium is the only Group 1 metal to form a nitride, Li_3N. The Group 2 metals all form nitrides.

All four of these properties can be attributed to the significant difference in ionic radius between the large "typical" Group 1 metals and the significantly smaller lithium ion (Table 11.1). For example, it is only the larger low-charge-density cations that can stabilize the large low-charge anions, such as hydrogen carbonate, in a crystal lattice. Lithium ion, being more the size of a Group 2 metal, cannot. Instead, formation of the higher lattice energy carbonate compound will be energetically preferred. The opposite argument can be used with the formation of ionic nitrides (and carbides, see in the following): that only a small higher charge-density cation can stabilize the high-charge anion.

Specific Resemblance of Lithium to Magnesium

The best examples of resemblance specifically between the chemistry of lithium and that of magnesium are

- The only tricarbides(4–), C_3^{4-}, of the Group 1 and 2 elements are Li_4C_3 and Mg_2C_3.
- Many lithium salts exhibit a high degree of covalency in their bonding as do those of magnesium.
- Lithium forms organometallic compounds similar to those of magnesium.
- Lithium is believed to occupy the same receptor site as magnesium in the treatment of bipolar disorder [20].

As with the nitrides, the formation of tricarbides can be interpreted in terms of stabilization by higher charge-density cations. The covalent behavior can also be explained as a result of lithium and magnesium being higher charge-density cations than the other members of their respective group.

Mackinnon [21] has pointed out that some other claimed evidence for the diagonality of these two elements is fallacious. For example, heating lithium nitrate gives lithium nitrite and oxygen gas (like the other Group 1 elements), not lithium oxide, nitrogen dioxide, and oxygen (like the Group 2 elements), contrary to some sources.

Does Li–Mg Isodiagonality Extend to Scandium? ... and Beyond?

Though diagonality has traditionally been considered a unique property of the early elements of the 2nd Period and

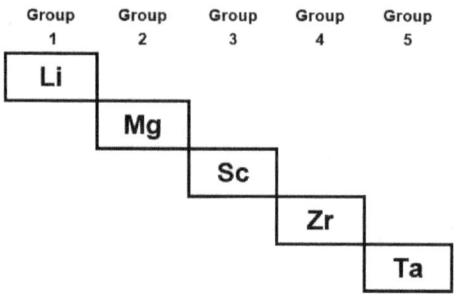

Figure 11.4 The first diagonal series.

3rd Period, there have been suggestions that diagonality extends into subsequent periods (as will be discussed in the following). The first complete diagonal series is shown in Figure 11.4.

As an example of isodiagonality extension, scandium, too, forms an insoluble fluoride, carbonate, and phosphate. Uniquely, scandium is the only other metal to form a carbide containing the tricarbide(4−) ion, though the compound, Sc_3C_4, also contains carbide(4−) and dicarbide(2−) ions within the same lattice structure. Also, in one of the few ores of scandium, *jervisite*, the same lattice site is occupied by scandium and magnesium: $(Na,Ca,Fe(II))(Sc,Mg,Fe(II))Si_2O_6$.

Scandium, in turn, has a resemblance to zirconium and thence to tantalum. For example, scandium forms $[Sc_6Cl_{12}]^{3-}$ clusters, while zirconium and tantalum (and also niobium) form related $[M_6Cl_{12}]$ cluster species.

Isodiagonality of Beryllium and Aluminum

The second pair to be examined here, that of beryllium and aluminum, has more common features unique to the diagonality. It is of relevance that, in the topological study of the chemical elements by Restrepo *et al.* [22], beryllium and aluminum were the only pair for which isodiagonal similarities exceed Group resemblances. This pair is the only diagonality example mentioned by House [23]. In addition to the other evidence that is widely cited, he refers to the two ions being particularly toxic — perhaps indicating a common biochemical bonding site.

Beryllium has a specific similarity to aluminum (and, to gallium) in terms of its aqueous (ionic) chemistry. Feinstein commented upon the similarities between the two elements in the context of analysis procedures. One example he gave was [24]:

> ... *the spectrophotometric method for beryllium or aluminum using the ammonium salt of aurin tricarboxylic acid.*

The similarity is particularly apparent when the Pourbaix (Eh–pH) diagrams [25] of the respective elements are compared (Table 11.2). The lower coordination number of the beryllium cation in acid solution may be explained as Be^{2+} being physically too small to accommodate six surrounding water molecules at a bonding distance.

In terms of compounds, there are several similarities:

- Beryllium and aluminum form carbides containing the carbide(4–), C^{4-} ion, both Be_2C and Al_4C_3 reacting with water to producing methane.

Table 11.2 A comparison of aqueous beryllium and aluminum species

	Acidic	Mid-pH Range	Basic
Beryllium	$[Be(OH_2)_4]^{2+}(aq)$	$Be(OH)_2(s)$	$[Be(OH)_4]^{2-}(aq)$
Aluminum	$[Al(OH_2)_6]^{3+}(aq)$	$Al(OH)_3(s)$	$[Al(OH)_4]^{-}(aq)$

- They form dimeric chlorides containing pairs of chlorine bridging atoms: $ClBeCl_2BeCl$ and $Cl_2AlCl_2AlCl_2$.
- The two elements form methyl organometallics, $Be(CH_3)_2$ and $Al(CH_3)_3$, with bridging CH_3 groups. Both compounds are spontaneously flammable in air and are explosively hydrolyzed by water.

Does Be–Al Isodiagonality Extend to Germanium? ... or to Titanium?

Roesky [26] has suggested that the diagonal relationship in this series continues to germanium. This proposal came as a result of his work on organometallic compounds of aluminum and attempts to synthesize germanium analogs.

However, Habashi [27] has pointed out that the chemistry of the aluminum ion more resembles that of the Group 3 elements rather than that of the lower members of Group 13 (see Chapter 9). Following from this proposal, the next member of the diagonal series should be considered as titanium. Titanium, like aluminum, is a low-density metal that reacts with the oxygen in air to form a tenacious protective layer of oxide to prevent further corrosion. One example of chemical similarities is that, for both aluminum and titanium, the fluorides are hexacoordinate species while the other halides are low-melting tetrahedrally coordinated species, such as Al_2Cl_6 and $TiCl_4$, which are hydrolyzed by water. The second complete diagonal series is shown in Figure 11.5.

Up to this point, similarities have been considered among valence-isoelectronic series, such as Be^{2+}, Al^{3+}, and Ti^{4+}. However, titanium also readily forms an ion Ti^{3+}. Thus, in this (and the later example of vanadium and molybdenum), matching compounds in the same oxidation can be considered. As an example here, titanium, like aluminum, forms alums such as $CsTi(SO_4)_2 \cdot 12H_2O$, analogous to $CsAl(SO_4)_2 \cdot 12H_2O$.

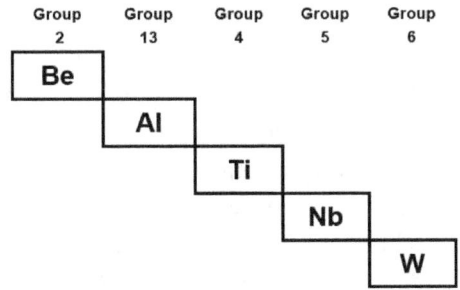

Figure 11.5 The second diagonal series.

Does Be–Al Isodiagonality Extend to Niobium? and to Tungsten?

There is some evidence for continuation of isodiagonality to niobium and tungsten. For example, both titanium and niobium form mixed metal oxides with a perovskite structure, such as $CaTiO_3$ and $LiNbO_3$, while tungsten(VI) oxide can adopt an analogous perovskite-like structure [28]. A key point to establishing the validity of an isodiagonal relationship is that of finding species unique to that linkage. Greenwood and Earnshaw have commented upon the relationship between the nonstoichiometric "bronzes" formed by titanium, niobium, and tungsten [29]. These compounds are characterized by very high electrical conductivities and intense colors. However, the diagonal similarities are better considered as two separate series: beryllium–aluminum and titanium–niobium–tungsten.

Isodiagonality of Boron and Silicon

A comparison of boron and silicon is the third common example of isodiagonality. This case is very different from the two other examples, for the chemistry of both elements involves essentially covalent bonding. Some of the similarities are listed here:

- Boron forms a solid acidic oxide, B_2O_3, like that of silicon, SiO_2. The oxide of aluminum is amphoteric.
- Boric acid, H_3BO_3, is a very weak acid as is silicic acid, H_4SiO_4. Boric acid bears no resemblance to the amphoteric aluminum hydroxide, $Al(OH)_3$.
- There are numerous polymeric borates and silicates that are constructed in similar ways, using shared oxygen atoms.
- Boron forms a range of flammable, gaseous hydrides, just as silicon does. There is only one aluminum hydride and that is a solid.

These similarities relate to covalent bonding. Unlike the explanations for the previous two pairs, there can be no justification in terms of cation charge density. Table 11.3 shows the element-oxygen single bond energy. Thus, it can be argued that it is the exceptional strength of the boron and silicon single bonds to oxygen that accounts for many of the diagonal similarities. This explanation would include the flammability of the hydrides to give the respective oxides.

There is more recent interest in diagonal relationships. A boron–silicon diagonal relationship has been claimed for a series of osmium organometallic compounds with boryl and corresponding dihydride silyl ligands [30]. While another diagonal connection is that boron and silicon are among the species to form heteropoly ions adopting Keggin structures [31].

Table 11.3 Element-oxygen single bond energy (kJ·mol⁻¹)

Group 13	Group 14	Group 15
Boron	Carbon	Nitrogen
536	358	201
—	Silicon	Phosphorus
	452	335

Isodiagonality of Carbon and Phosphorus

Inorganic chemists have seen the isodiagonal relationship primarily in terms of the three pairs earlier. A landmark advance was the synthesis of a ferrocene-like molecule in which the ring carbon atoms were replaced by phosphorus. Not only did this molecule resemble ferrocene, but it underwent a substitution reaction in a similar manner to ferrocene itself [32].

In recent years, organic chemists have been using the term "diagonal relationship" in the context of organophosphorus compounds. By 1998, there had been so much organophosphorus chemistry reported that a review book appeared with the title: *Phosphorus: The Carbon Copy* [33]. Some of the research has continued to cite the "diagonal relationship" as the underlying premise. For example, research published in 2000 describes the phospha-Wittig reaction [34]. Subsequently, the diagonal relationship was cited in the context of phosphorus-containing heterocycles [35] and in a review of poly-phospho-cation species [36].

Isodiagonality of Nitrogen and Sulfur

Greenwood and Earnshaw [37] suggested that similarities in charge densities and electronegativities between nitrogen and sulfur would lead to a diagonal relationship between these two elements. As evidence, they used the extensive range of cyclo binary sulfur nitrides. Greenwood and Earnshaw contended that the large number of permutations, or their interchangeability, indicated strong similarities between the two elements. Of particular relevance is the aromatic nature of the cyclo-$S_3N_2^{2+}$ ion, suggesting the closeness in electronic energies of the two component atoms [38].

Isodiagonality of Vanadium and Molybdenum

Mitchell proposed a link between the chemistry of vanadium and molybdenum in 1974 [39]. At the time, he argued that, in its chemistry, vanadium more closely resembled molybdenum and tungsten than the other members of its own group. Mitchell also noted similarities between the chemistry of molybdenum and rhenium.

In 1986, it was reported that the nitrogenase enzyme was sometimes found with vanadium instead of molybdenum [40]. Subsequently, the vanadium–molybdenum link has also become of interest in the context of other enzymes that can utilize vanadium in place of molybdenum. Rehder proposed that, in early geological times, vanadium was more widely used than molybdenum in enzyme processes [41]. This complete diagonal series is shown in Figure 11.6.

In their aqueous chemistry, there are far more similarities of vanadium and molybdenum than of vanadium and niobium. At high pH, soluble vanadate ion, VO_4^{3-}, and

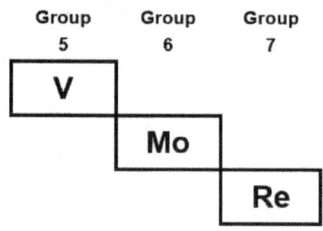

Figure 11.6 The Group 5–7 diagonal series.

Table 11.4 A comparison of aqueous vanadium and molybdenum species under oxidizing conditions

	Acidic		Mid-pH Range	Basic
Vanadium	$VO_2^+(aq)$		Soluble isopolyvanadates	$VO_4^{3-}(aq)$
Molybdenum	$MoO_2^{2+}(aq)$	$MoO_3(aq)$	Soluble isopolymolybdates	$MoO_4^{2-}(aq)$

molybdate ion, MoO_4^{2-} predominate. Then at intermediate pH, they both form soluble isopoly-oxo-anions. At low pH, vanadium forms a soluble *cis*-dioxo-cation, VO_2^+, while molybdenum forms an insoluble trioxide at low pH and the corresponding *cis*-dioxo-cation, MoO_2^{2+}, at very low pH (Table 11.4).

There are also analogous phosphorus-containing heteropoly-oxo-anions, including $[PV_{14}O_{42}]^{9-}$ and $[PMo_{12}O_{40}]^{3-}$. By contrast, vanadium does not resemble niobium in its predominant species. For example, niobium forms insoluble niobium(V) oxide over much of the oxidized pH range [24]. Among other similarities between vanadium and molybdenum, vanadium(IV) and molybdenum(IV) form disulfides, VS_2 and MoS_2, with matching layer structures.

Does V–Mo Isodiagonality Extend to Rhenium?

Mitchell [39] alluded to a diagonal relationship between molybdenum and rhenium. One of several such similarities is the formation of analogous valence-isoelectronic chlorodimers containing quadruple bonds and eclipsed chlorine atoms: $[Mo_2Cl_8]^{4-}$ and $[Re_2Cl_8]^{2-}$. There are also some features that encompass all three of the diagonal series, such as reduction under the same very acid conditions to the identical oxidation state: V^{3+}, Mo^{3+}, and Re^{3+}. In addition, all three elements form corresponding valence-isoelectronic tetrathioanions: VS_4^{3-}; MoS_4^{2-}; and ReS_4^- [42].

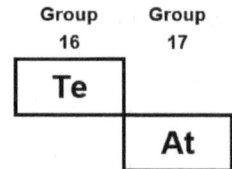

Figure 11.7 The Group 16–17 lower diagonal series.

Table 11.5 A comparison of aqueous tellurium and astatine species under oxidizing conditions

	Acidic	Mid-pH Range	Basic
Tellurium	H_6TeO_6	$H_5TeO_6^-$	$H_4TeO_6^{2-}$
Astatine	H_5AtO_6	$H_4AtO_6^-$	$H_3AtO_6^{2-}$

Isodiagonality of Tellurium and Astatine

It is likely that other examples of isodiagonality remain to be explored. In fact, isodiagonality seems to occur in many parts of the Periodic Table. For example, there are similarities between the chemistry of tellurium(VI) and astatine(VII) (Figure 11.7).

These resemblances are not just in formulas, but also in the pH dependence of the oxo-anions (Table 11.5).

Evidence-Based Isodiagonality

It is probable that at least one similarity could be found between any pair of elements. Thus, it is important to define criteria that can be used to establish an isodiagonal relationship as a real phenomenon. There seems to be three separate ways that can be used individually — or preferably in combination — to define elements as having an isodiagonal relationship.

- That for several compounds, there are diagonal similarities in physical and/or chemical properties. This is the classic definition used for the Li–Mg, Be–Al, and B–Si relationships.
- That very unusual chemical structures are shared in a diagonal pattern. Examples of these would be the tricarbide(4–) found for the Li–Mg pair and the *cis*-dioxocations of the V–Mo pair.
- That in a few cases, there is an isodiagonal relationship between two elements that share the same biological function or occupy the same enzyme site. Examples of this would be the Li–Mg pair and the V–Mo pair, both discussed earlier.

Evidence of isodiagonality is usually found by considering valence-isoelectronic species, but for elements that have more than one oxidation state, similarities can sometimes be found for a common oxidation state.

Commentary

It would be nice to discover an all-encompassing explanation for isodiagonal relationships. As they are an upper-left–lower-right pattern, electronegativity might be used to account for some of the resemblances, while similarities in ionic radius and/or charge density might explain others. However, there seems to be no simple general "explanation" for isodiagonality. From the perspective of this series, however, the more important task is to define and clarify the isodiagonal relationship as one of the valid linkages among the chemical elements — and not just limited to the upper-left corner of the Periodic Table.

References

1. A. J. Ihde, *The Development of Modern Chemistry*, Dover, New York, NY, 247 (1984).
2. S. J. French, "Warping the Periodic Table," *J. Chem. Educ.* **14**, 571–573 (1937).
3. R. L. Rich, "A Taxonomy of Relationships among the Elements," *J. Chem. Educ.* **68**, 828–829 (1991).
4. R. L. Rich, "Are Some Elements More Equal than Others?" *J. Chem. Educ.* **82**, 1761–1763 (2005).
5. P. P. Edwards and M. J. Sienko, "On the Occurrence of Metallic Character in the Periodic Table of Elements," *J. Chem. Educ.* **60**, 691–696 (1983).
6. S. J. Hawkes, "Semimetallicity?" *J. Chem. Educ.* **78**, 1686–1687 (2001).
7. D. M. P. Mingos, *Essential Trends in Inorganic Chemistry*, Oxford University Press, Oxford, 207 (1998).
8. G. H. Cartledge, "Studies on the Periodic System: II the Ionic Potential and Related Properties," *J. Am. Chem. Soc.* **50**, 2863–2872 (1928).
9. T. P. Hanusa, "ReExamining the Diagonal Relationships," *J. Chem. Educ.* **64**, 686–687 (1987).
10. R. J. Puddephatt and P. K. Monaghan, *The Periodic Table of the Elements*, 2nd ed., Clarendon Press, Oxford, 64–67 (1986).
11. J. D. Lee, *Concise Inorganic Chemistry*, 5th ed., Chapman & Hall, London, 189–190, 310–311, 353–354, 389 (1999).
12. R. B. King, *Inorganic Chemistry of the Main Group Elements*, Wiley, New York, NY, 200 (1995).
13. C. E. Housecraft and A. G. Sharpe, *Inorganic Chemistry*, 3rd ed., Pearson Education, Harlow, 321–322 (2008).
14. G. Rayner-Canham and T. Overton, *Descriptive Inorganic Chemistry*, 6th ed., W. H. Freeman, New York, NY, 227–229, A13–A15 (2014).

15. J. W. Eichinger, Jr. "The Cesium Ratio, a Simple Charge Density Calculation," *J. Chem. Educ.* **39**(10), 536–537 (1962).
16. G. E. Rogers, *Descriptive Inorganic, Coordination, and Solid-State Chemistry*, 3rd ed., Brooks/Cole, Belmont, CA, 239–242, 328–329, 406 (2015).
17. R. D. Shannon, "Revised Effective Ionic Radii and Systematic Studies of Interatomic Distances in Halides and Chalcogenides," *Acta Cryst. Sect. A* **32**(5), 751–767 (1976).
18. A. L. Allred and E. G. Rochow, "A Scale of Electronegativity Based on Electrostatic Force," *J. Inorg. Nucl. Chem.*, **5**(4), 264–268 (1958).
19. N. N. Greenwood and A. Earnshaw, *Chemistry of the Elements*, 2nd ed., Butterworth-Heinemann, Oxford, 76 (1997).
20. T. M. Brown, A. T. Dronsfield, and P. M. Ellis, "Li-Mg: A Life-Saving Relationship," *Educ. Chem.* **34**, 72–74 (1997).
21. A. Mackinnon, "Are Our Relationships Diagonal?" *School Sci. Rev.* **61**, 165 (1979).
22. G. Restrepo *et al.*, "Topological Study of the Periodic System," *J. Chem. Inf. Comput. Sci.* **44**, 68–75 (2004).
23. J. E. House, *Inorganic Chemistry*, Academic Press, Burlington, MA, 370–371 (2008).
24. H. I. Feinstein, "Diagonal Relationships-Descriptive or Theoretical?" *J. Chem. Educ.* **61**(2), 128 (1984).
25. G. K. Schweizer and L. L. Pesterfield, *Aqueous Chemistry of the Elements*, Oxford University Press, Oxford (2010).
26. H. W. Roesky, "Preparation of Fluorine Compounds of Groups 13 and 14: A Study Case for the Diagonal Relationship of Aluminum and Germanium," *J. Fluor. Chem.* **125**, 1765–1769 (2004).
27. F. Habashi, "A New Look at the Periodic Table," *Interdiscipl. Sci. Rev.* **12**, 53–60 (1997).
28. A. Magnéli, "Some Aspects of the Crystal Chemistry of Oxygen Compounds of Molybdenum and Tungsten Containing Structural Elements of ReO_3 or Perovskite Type," *J. Inorg. Nucl. Chem.* **2**(5–6), 330–339 (1956).
29. N. N. Greenwood and A. Earnshaw, *Chemistry of the Elements*, 2nd ed., Butterworth-Heinemann, Oxford, 987 (1997).
30. M. L. Buil *et al.*, "Cationic Dihydride Boryl and Dihydride Silyl Osmium(IV) NHC Complexes: A Marked Diagonal Relationship," *Organometallics* **32**(9), 2744–2752 (2013).

31. P. G. Rasmussen, "Some Perspectives on Heteropoly Ion Chemistry," *J. Chem. Educ.* **44**(5), 277–279 (1967).

32. R. Bartsch, P. B. Hitchcock, and J. F. Nixon, "First Structural Characterisation of Penta- and Hexa-Phosphorus Analogues of Ferrocene. Synthesis, Crystal and Molecular Structure of the Air-Stable, Sublimable Iron Sandwich Compounds [Fe(η^5-$C_2R_2P_3)_2$], and [Fe(η^5-$C_3R_3P_2$)(η^5-$C_2R_2P_3$)](R = But)," *J. Chem. Soc. Chem. Commun.* 1146–1148 (1987).

33. K. B. Dillon, F. Mathey, and J. F. Nixon, *Phosphorus: The Carbon Copy: From Organophosphorus to Phospha-Organic Chemistry*, Wiley, Chichester (1998).

34. S. Shah and J. D. Protasiewicz, "'Phospha-Variations' on the Themes of Staudinger and Wittig: Phosphorus Analogs of Wittig Reagents," *Coord. Chem. Rev.* **210**, 181–200 (2000).

35. A. Modelli, B. Hajgató, J. F. Nixon, and L. Nyulászi, "Anionic States of Six-Membered Aromatic Phosphorus Heterocycles as Studied by Electron Transmission Spectroscopy and *ab initio* Methods," *J. Phys. Chem. A* **108**, 7440–7447 (2004).

36. C. A. Dyker and N. Burford, "Catena-Phosphorus Cations," *Chem. Asian J.* **3**, 28–36 (2008)

37. N. N. Greenwood and A. Earnshaw, *Chemistry of the Elements*, 2nd ed., Butterworth-Heinemann, Oxford, 722 (1997).

38. B. M. Gimarc and D. S. Warren, "Molecular Structures, Stabilities, and Electronic States of the Planar Rings $S_3N_2^+$ and $S_3N_2^{2+}$," *Inorg. Chem.* **30**, 3276–3280 (1991).

39. P. C. H. Mitchell, "The Chemistry and Uses of Molybdenum: Introductory Lecture," *J. Less-Common Metals* **36**, 3–11 (1974).

40. R. L. Robson *et al.*, "The Alternative Nitrogenase of *Azotobacter chroococcum* is a Vanadium Enzyme," *Nature* **322**, 388–390 (1986).

41. D. Rehder, "Is Vanadium a More Versatile Target in the Activity of Primordial Life Forms than Hitherto Anticipated," *Organic Biomol. Chem.* **6**, 957–964 (2008).

42. R. H. Petit *et al.*, "Magnetic Circular Dichroism and Absorption Spectra of d^0 Tetrahedral Oxyanions and Thioanions: MoS_4^{2-}, MoO_4^{2-}, WS_4^{2-}, ReS_4^-, VS_4^{3-}, VO_4^{3-} and OsO_4," *Mol. Phys.* **27**(5), 1373–1384 (1974).

Chapter 12

Lanthanoids, Group 3, and Their Connections

The short form of the conventional Periodic Table may be convenient and compact, but it results in a marginalization of the lanthanoids (and of the actinoids, covered in Chapter 13). Not only do they appear to be the orphan elements, but in undergraduate chemistry courses, if they are mentioned at all, they are dismissed in the closing lecture(s) of the course as being boring and of little interest. Yet this is not the case. In addition to some trends, there are also interesting and curious exceptions. Lanthanoid chemistry is most definitely worthy of study.

In Chapter 5, the rare earth metals were defined as the lanthanoids (see in the following) plus the two earlier Group 3 elements of scandium and yttrium. Cotton has referred to these two elements as "The Misfits" of the Periodic Table [1] and these two elements will be discussed first.

Yttrium and Scandium

In the debate on the 6th Period and 7th Period members of Group 3 (see Chapter 4), the 4th Period and 5th Period members of the Group are often overlooked. In Chapter 8, scandium and yttrium were rejected from membership of the transition metals. They have found a home in this book by joining the lanthanoids as part of the rare earth metal grouping.

Yttrium

The yttrium 3+ ion resembles the lanthanoid 3+ ions so closely that it is best regarded as a later lanthanoid. For example, yttrium's standard reduction potential is −2.37 V compared with −2.37 V for lanthanum and −2.30 V for lutetium.

Also, the yttrium 3+ ion has an ionic radius of 90.0 pm compared with 90.1 pm for holmium.

In the context of its chemistry, the yttrium(III) halides are isostructural with all the lanthanoid(III) halides from dysprosium to lutetium. The yttrium(III) ion exists as an octahydrate is aqueous solution, $[Y(OH_2)_8]^{3+}$, as do many of the lanthanoids. The major interest in ytterbium has been for yttrium oxosulfide, Y_2O_2S, doped with later lanthanoid(III) ions as long-lasting phosphors [2].

Though yttrium seems "more comfortable" with the latter lanthanoids, curiously, yttrium in minerals is usually associated with the earlier lanthanoids. Two examples of these are *bastnäsite*, $(Ce,La,Y)CO_3F$, and *gadolinite* (which contains no gadolinium), $(Ce,La,Nd,Y)_2FeBe_2Si_2O_{10}$.

Though yttrium does seem to be predominantly lanthanoid-like, there seems to be a significant resemblance in chemistry to its $(n + 10)$ Group 13, analog, indium (see Chapter 9). Interestingly, research on inorganic polymers has also found some similarities of yttrium and indium [3].

Scandium

While yttrium behaves unambiguously like a lanthanoid, the much smaller scandium(III) ion (74.5 pm) has resemblances to — and differences from — both transition metals and lanthanoids. In fact, Cotton's term of "misfit" does indeed apply to this element. For example, like the transition metal ions from titanium(III) to cobalt(III), scandium(III) forms a three-coordinate silylamide, $Sc[N(Si(CH_3)_3)_2]_3$. However, while the transition metal complexes are planar, that of scandium is pyramidal. Despite the difference in ion size, scandium can be found with yttrium in such ores as *thortveitite*, $(Sc,Y)_2Si_2O_7$.

In Chapter 9, the several similarities of scandium to aluminum were discussed in the context of the (n) and ($n + 10$) relationship. The major use for scandium is in specialty alloys of aluminum, the Al_3Sc micrograins imparting additional strength to the aluminum metal [4]. In fact, in some respects, scandium is closer in chemistry to aluminum than to any other element.

The 4f Elements

When the first rare earth elements were discovered, the question arose as to where they could be fitted in the eight-group Periodic Table. At the time, it was believed that the eight-group framework was the key "set in stone" feature of the Periodic Table: the elements themselves were the problem. It was not until 1882 that Bailey concluded that the known rare earth elements were not members of any of the eight Groups [5]. Then Bassett, in 1892, proposed that the lanthanoid elements (as we now call them) form their own series (and that the known actinoids formed a matching series). The key to the understanding of the lanthanoids came in 1921, when Bury postulated that they corresponded to the filling of the 4f orbitals.

Progressing to the present, what of their chemistry? Pimentel and Sprately summed up the opinion of most chemists to the chemistry of the 4f elements [6]:

> Lanthanum has only one important oxidation state in aqueous solution, the +3 state. With few exceptions, this tells the whole boring story about the other lanthanides.

Yet the predominance of the +3 state is one of the very interesting things about the 4f elements. Nowhere else in the Periodic Table is it possible to study a sequential series of elements all in the same oxidation state. And there are many other interesting aspects to their chemistry as will be covered in the following.

Books specifically on the 4f elements (or the "rare earth" elements, which also encompass Group 3 elements — see Chapter 4) are very rare. One volume was part of a series on each segment of the Periodic Table [7]. An undergraduate textbook claiming to be on the d-block and f-block elements contained very little on the f-block, instead being almost entirely on the d-block [8]. Though the text by Cotton (mentioned earlier) [1] was comprehensive, it did not look beyond the confines of the f-block for relationships, nor did the similar book by Aspinall [9]. There has also been a compilation of studies of the coordination chemistry of the rare earth elements [10].

Yet in the research world, lanthanoids have been a burgeoning field. The metals and their alloys have become indispensable as magnetic materials [11]. In the 1990s, the primary interest in their compounds was as reagents in organic synthesis [12, 13]. Now the focus has become on the luminescent properties of the lanthanoid ions and their applications [14–16]. There is also an interest in a group of extremophile aerobic methanotrophs that require one of the early lanthanoids (lanthanum, cerium, praseodymium, or neodymium) for their metabolic pathways [17].

The Lanthanoids

To begin, as is recognized by the International Union of Pure and Applied Chemistry (IUPAC), the correct term is "lanthanoids" [18]. The ending "-ide" is accepted throughout chemistry nomenclature as referring to a negative ion,

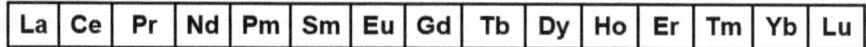

Figure 12.1 The lanthanoid elements as defined in this chapter.

as in "oxide" or "sulfide." The commonly accepted definition of a lanthanoid is therefore:

> A **lanthanoid** *is any of the series of 15 consecutive chemical elements in the Periodic Table from lanthanum to lutetium (Figure 12.1).*

Properties of the Elements

There is the inference that, for a particular property of the lanthanoid elements, a linear or smooth curve plot results. This is not necessarily true. Cater showed that for several parameters of high-temperature lanthanoid chemistry, the plots are much more uneven [19]. This discovery was revisited by Johnson, who came to the following conclusion [20]:

> ... *the lanthanide elements behave similarly in reactions in which the 4f electrons are conserved, and very differently in reactions in which the number of 4f electrons changes.*

Laing showed there were significant deviations from linearity for lanthanoid melting points. Though there is a general trend of increasing melting points of the lanthanoids with increasing atomic number, there are two exceptions to the rule: europium and ytterbium, which both have melting points well below that of the trend. He ascribed these anomalies to much weaker metallic bonds for these two elements [21].

In addition, Laing noted that the densities of the lanthanoid metals followed an even more linear relationship (see Figure 12.2), though again, with the exception of europium and ytterbium [21]. He accounted for these two deviations in terms of the electron-sea model of metallic bonding. For all other lanthanoids, the intermetallic forces involved

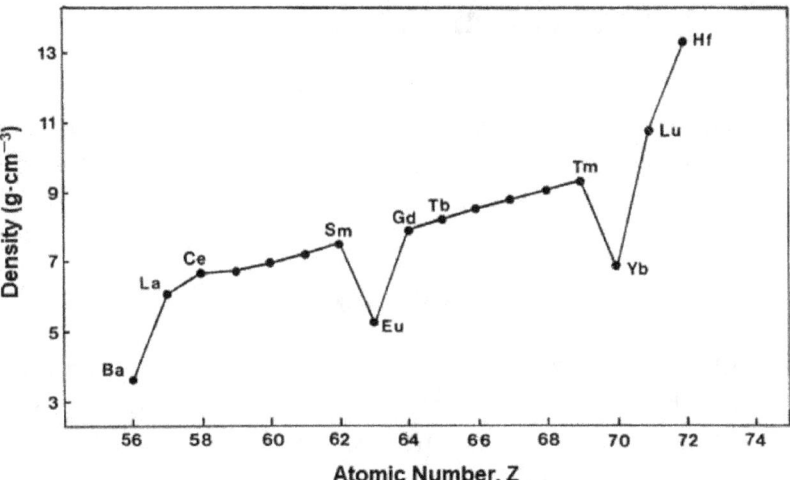

Figure 12.2 Densities of the elements from atomic number 56 to 72 (adapted from Ref. [20]).

3+ ions and the intervening three "roaming" valence electrons. Laing argued that as europium and ytterbium favored the 2+ state (see in the following), then the intermetallic forces between the (theoretical) 2+ ions and two "roaming" electrons would be significantly less.

However, though the electron-sea model can provide a very simplistic idea of metal behavior, it is incapable of being given any quantitative or even semiquantitative validity. Without going into the sophistication of band theory, the alternative is the soft-sphere model. Lang (not to be confused with Laing) has applied the soft-sphere model to the lanthanoid metals and showed that it is a good fit [22]. Nevertheless, it is more of a justification than an explanation, in that the calculation requires knowledge and use of the crystal packing factor. It does not explain why europium uniquely has the body-centered cubic packing rather than the more compact (and therefore denser) packing arrangements of the other lanthanoids.

Ion Charges of the Lanthanoids

The 3+ state predominates for all of the lanthanoids [23]. It is the 3+ oxidation state consistency that gives such a useful comparison across the series.

The Lanthanoid Contraction

Of importance is the 3+ ionic radii of the lanthanoids. As can be seen from Figure 12.3, the ionic radius decreases, almost linearly, from lanthanum to lutetium. This decrease is known as the lanthanoid contraction (or more commonly as the "lanthanide contraction") [24]. The effect was first recognized and named by the Norwegian geochemist, Victor Goldschmidt. The contraction, defined in the following, is an important aspect of lanthanoid and, even post-lanthanoid chemistry [25].

> The **lanthanoid contraction** is the greater-than-expected decrease in ionic radii of the elements from lanthanum to lutetium, which results in smaller than otherwise expected ionic radii for the subsequent elements commencing with hafnium.

The lanthanoid contraction can be explained as follows [26]. It is the inner, filled, $5s^2 5p^6$ electron "layer" that defines

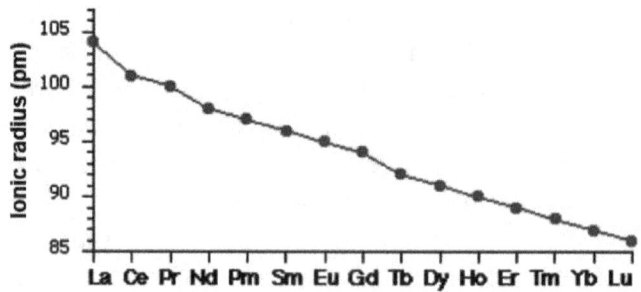

Figure 12.3 The lanthanoid contraction for the 3+ ions.

the ionic radius. The 4f electrons contribute little to the shielding. Thus, as the nuclear charge increases, there is a contraction of the $5s^25p^6$ orbitals causing a radius reduction of the ions.

It is a common assumption that, having given it a special name, the lanthanoid contraction is greater than other contractions across periods. This is not true, as Lloyd has pointed out [27]. In fact, the decrease in radius from lanthanum(3+) to lutetium(3+) of 117 pm to 100 pm is less than that from calcium(2+) to zinc(2+) of 114 pm to 88 pm. From one ion to its neighbor, the average individual lanthanoid contraction is also less than that from scandium(3+) to gallium(3+) of 89 pm to 76 pm.

Post-Lanthanoid Effect of the Atom and Ion Contraction

As described in Chapter 8, a crucial consequence of the 14-element contraction is that the 6th Period transition metal series are of almost the same atom and ionic radius as their 5th Period analogue. As an example, the pair of zirconium and hafnium can be considered. Comparing these two elements, the number of protons (and number of electrons) has increased from 40 to 78. Yet the atomic radius decreases from 159 pm to 156 pm. Similarly, there is a small decrease, not an increase, in ionic radii from 86 pm (Zr^{4+}) to 85 pm (Hf^{4+}).

Had the lanthanoids not intervened, it is possible to make a rough estimate of the Hf^{4+} ion radius. This can be done by comparing the ionic radius of the 3+ ion preceding the lanthanoids, lanthanum (117 pm) with the element above it, yttrium (104 pm). Using approximately the same difference, without the lanthanoid contraction, Hf^{4+} ion would be expected to be about 10 pm larger than zirconium, instead of 1 pm smaller.

Other Lanthanoid Oxidation States

It is the variations in oxidation states from the +3 "norm," which have provided the most interest [26]. When plotting out known oxidation states for elements, the question arises as to how extreme the conditions, or how unusual the ligands, that have been used in order to stabilize a specific oxidation state. Table 12.1 identifies those oxidation states that have a significant existence for simple compounds. In contradiction to this generality, praseodymium(V) is included (in parenthesis) as it is of intrinsic interest in the context of $4f^0$ configurations. It can be seen that the empty; half-filled; and filled f electron energy state plays a significant — but not exclusive role — in determining which other oxidation state(s) are feasible for a specific lanthanoid.

The $4f^0$ oxidation state is the expected state for lanthanum. For cerium, Ce(IV) is a common state, though highly oxidizing. In the next section, we will see that cerium(IV) has many similarities to members of Group 4. The oxidation of cerium(III) to cerium(IV) has relevance to geochemistry. In oxidizing waters, cerium is deposited as insoluble cerium(IV) oxide. This is one parameter by which the redox condition of ancient seas and oceans can be determined

Table 12.1   The 4f electron configurations corresponding to the common ion charges (adapted from Ref. [26])

Element / Charge	La	Ce	Pr	Nd	Pm	Sm	Eu	Gd	Tb	Dy	Ho	Er	Tm	Yb	Lu
+5			$(4f^0)$												
+4		$4f^0$	$4f^1$						$(4f^7)$						
+3	$4f^0$	$4f^1$	$4f^2$	$4f^3$	$4f^4$	$4f^5$	$4f^6$	$4f^7$	$4f^8$	$4f^9$	$4f^{10}$	$4f^{11}$	$4f^{12}$	$4f^{13}$	$4f^{14}$
+2						$4f^4$	$4f^7$							$4f^{14}$	

[29]. The cerium enrichment (as cerium(IV) oxide) compared with the other lanthanoids is known as the "positive" cerium anomaly.

Though an empty f shell would seem to be an obvious possibility for praseodymium, it seems that Pr(V) is, in fact, not a particularly favored oxidation state for the element. The only species obtained by the date of writing has been the ion $[PrO_2]^+$ under very low-temperature noble gas matrix isolation [28].

Several of the lanthanoids exhibit the +2 oxidation state [30], but it is only for europium and ytterbium that the +2 state is of major importance. In addition, terbium only "reluctantly" forms compounds in the +4 oxidation state [31], which is surprising considering it has a significantly lower 4th ionization energy [32].

As can be seen from Table 12.1, the progression: lanthanum(III); cerium(IV); and praseodymium(V) corresponds to the "empty-shell" $4f^0$ series, which would not be unexpected [33]. Similarly, ytterbium(II) and lutetium(III) correspond to the "full-shell" $4f^{14}$ configurations. The third isoelectronic set, europium(II); gadolinium(III); and terbium(IV) correspond to the "half-filled" $4f^7$ electron configuration. This so-called "stability of the half-filled shell" (see Chapter 2) is often discussed in the context of the main group elements [34] but it is also evident for the lanthanoids.

Restructuring the Lanthanoids

Though conventionally the lanthanoids are treated as a single continuous unit, attempts have been made to identify subcategories and possible rearrangements.

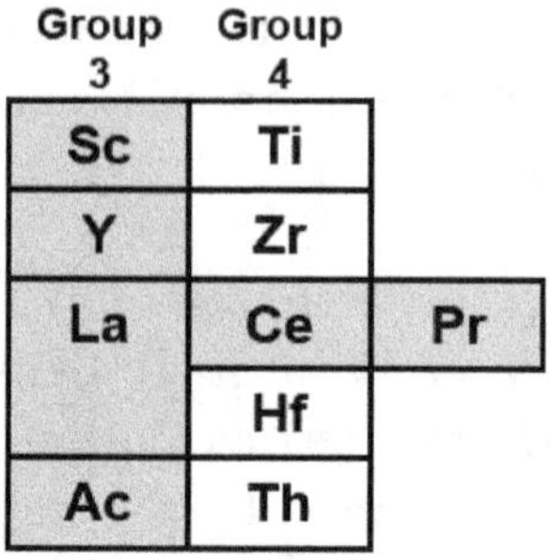

	Group 3	Group 4	
	Sc	Ti	
	Y	Zr	
	La	Ce	Pr
		Hf	
	Ac	Th	

Figure 12.4 Cerium as a member of Group 4 in addition to the lanthanoids (modified from Ref. [35]).

Cerium as a Member of Group 4

Johansson *et al.* have singled out cerium on the basis of its +4 oxidation state to be better considered as a member of Group 4 [35]. This assignment is shown in Figure 12.4.

The Stacked Lanthanoid Arrangement

In geochemistry, the rare earth elements are classified as "light" or "heavy." Of the lanthanoids, lanthanum to gadolinium are usually considered as "light" while terbium to lutetium are usually assigned as "heavy." More of the "light" lanthanoids are in the Earth's crust, while more of the "heavy" lanthanoids in the Earth's mantle. This distinction comes about through the variation in ionic radii and hence the crystal structures in which the ion will fit. The trend of ionic radii results from the lanthanide contraction described earlier.

It was Ternström in 1976, who first proposed "stacking" the two half series on the basis of physical and chemical properties as: Ce–Gd and Tb–Lu [36]. Laing elaborated upon this concept, focusing upon chemical resemblances

[21]. He noted (as stated earlier) that both europium and ytterbium form compounds in which they have a +2 oxidation state and therefore related to Group 2. Similarly, cerium commonly forms compounds in which it has the +4 oxidation state, and therefore should be associated with Group 4. Laing placed these three elements in their assigned groups, plus intervening lanthanum, gadolinium, and lutetium in Group 3. The other lanthanoids were then placed in order to complete each of the two subrows (Figure 12.5).

Though the early members fit, there is no evidence so far of any higher oxidation states for the later members of the lanthanoids. Laing subsequently changed his mind about the arrangement of the lanthanoids [37]. Instead, he devised a three-level sandwich that highlighted the +2 trio of [Ba–Eu–Yb], the +3 trio of [La–Gd–Lu], and the +4 trio of [Ce–Tb–Hf], with gadolinium being central, as shown in Figure 12.6.

Group 2	Group 3	Group 4	Group 5	Group 6	Group 7	Group 8
Ba	La	Ce	Pr	Nd	Pm	Sm
Eu	Gd	Tb	Dy	Ho	Er	Tm
Yb	Lu	Hf	Ta	W	Re	Os

Figure 12.5 The two-row lanthanoids according to Laing, with the surrounding elements shaded (modified from Ref. [21]).

38	39						56	57	58						39	40
Sr	Y						Ba	La	Ce						Y	Zr
56	57	58	59	60	61	62	63	64	65	66	67	68	69	70	71	72
Ba	La	Ce	Pr	Nd	Pm	Sm	Eu	Gd	Tb	Dy	Ho	Er	Tm	Yb	Lu	Hf
							70	71	72							
							Yb	Lu	Hf							

Figure 12.6 Laing's gadolinium-centered lanthanoid series (from Ref. [37]).

Dendrogram Restructuring

Horovitz and Sârbu used a cluster analysis to develop a similar double-row set [38]. The dendrogram that relied largely upon a variety of numerical values for each atom/ion is shown here (Figure 12.7).

From the structure of the dendrogram, Horovitz and Sârbut devised a segment of the Periodic Table. This arrangement (Figure 12.8) largely matches Laing's two-row lanthanoid pattern shown in Figure 12.5 (though Laing contested the differences [39]), except that the Eu–Yb pair are

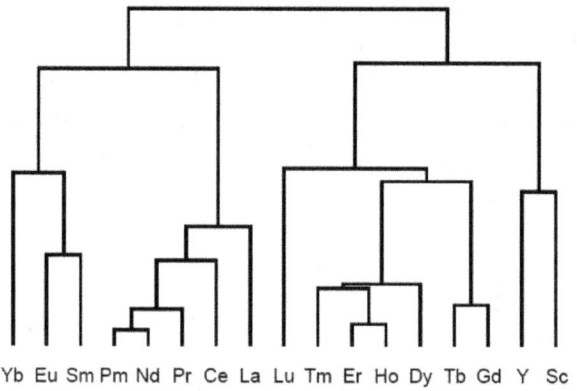

Figure 12.7 A dendrogram for the lanthanoids (adapted from Ref. [38]).

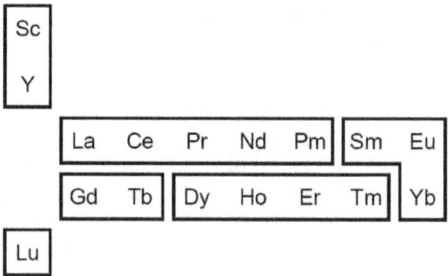

Figure 12.8 Lanthanoids arranged according to the cluster analysis of Horovitz and Sârbut (adapted from Ref. [38]).

shown at the right-hand, not left-hand, end. Noticeably, the two-row sequence is fragmented.

The clustering of samarium with europium and ytterbium is particularly significant. Together with europium and ytterbium, samarium is the only other lanthanoid to form stable 2+ compounds. In fact, samarium(II) iodide (Kagan's reagent) is an important mild reducing agent in organic chemistry [40].

The gap between promethium and samarium according to the cluster analysis is interesting in that it seems to be on the borderline of some ion-packing arrangements. For the lanthanoid(III) fluorides, from lanthanum to promethium, the crystal structure is based upon the nine-coordinate lanthanum(III) fluoride tricapped trigonal prism arrangement. Then from samarium to lutetium, the crystal structure is, by contrast, based upon the eight-coordinate yttrium(III) fluoride bicapped trigonal prism. Similarly, for the lanthanoid(III) iodides, from lanthanum to promethium, the crystal structure is based upon the eight-coordinate plutonium(III) bromide bicapped trigonal prism. Then from samarium to lutetium, the crystal structure is based upon the six-coordinate iron(III) chloride octahedral structure.

External Lanthanoid Relationships

Surprisingly, all the lanthanoid(III) (and yttrium(III)) ions have been shown to be essential cofactors in certain enzymes in methylotrophic bacteria. They bind 10^8 times more strongly to the ligand sites than calcium, which is the usual metal ion in such enzymes [41]. Nevertheless, it is with the two "unusual" oxidation states of lanthanoids for which there are interesting similarities with other elements.

Similarities of Europium(II) with Strontium (and Calcium)

The europium(II) ion behaves very similarly to an alkaline earth ion; for example, its carbonate, sulfate, and chromate are insoluble, as are those of the heavier alkaline earth metals. The ionic radius of europium(II) is actually very similar to that of strontium, and, as might be expected, several europium(II) and strontium compounds are isostructural.

In the study of lanthanoid-containing minerals, the proportion of europium can be significantly different from that of the other lanthanoids [42]. In anaerobic geothermal conditions, europium(III) can be reduced to europium(II). Then, though significantly larger than the calcium ion, europium(II) can replace calcium in minerals. This ion exchange process is known as the europium anomaly and it is said to be "positive" if europium is enriched with respect to the other lanthanoids and "negative" if it is depleted.

Similarities of Cerium(IV) with Zirconium(IV) and Hafnium(IV)

Whereas europium has a lower than normal oxidation state, cerium has a higher than normal oxidation state of +4. Cerium(IV) behaves like zirconium(IV) and hafnium(IV) of Group 4. For example, all three of these ions form insoluble fluorides and phosphates. The similarity can be seen from a comparison of acid–base behavior under strongly oxidizing conditions (Table 12.2).

The ready oxidation of cerium(III) to cerium(IV) has significant geochemical implications. For example, the similar ion sizes of cerium(IV) and zirconium(IV) resulted in

Table 12.2 Schematic of predominant species with pH under oxidizing conditions

	Very Acidic	Acidic	Basic	Very Basic
Zirconium	$Zr^{4+}(aq)$	$ZrO_2(s)$		
Hafnium	$Hf^{4+}(aq)$	$HfO_2(s)$		
Cerium	$Ce^{4+}(aq)$	$CeO_2(s)$		

incorporation of Ce^{4+} ions into zircons, key minerals in the context of the earliest Earth's rocks [43].

Commentary

It is time for the lanthanoids to have their "day in the Sun" and not be relegated to a passing mention — if anything at all — in chemistry courses. Apart from the fact that they do have interesting chemistry, their many applications in the real world require students to be aware of them.

References

1. S. Cotton, *Lanthanide and Actinide Chemistry: Inorganic Chemistry (A Textbook Series)*, Wiley, Chichester, England (2006).
2. H. Liu *et al.*, "One-Pot Solvothermal Synthesis of Singly Doped Eu^{3+} and Codoped Er^{3+},Yb^{3+} Heavy Rare Earth Oxysulfide Y_2O_2S Nano-Aggregates and their Luminescence Study," *RSC Adv.* **4**(100), 57048–57053 (2014).
3. S. P. Petrosyants, "Coordination Polymers of Indium, Scandium, and Yttrium" *Russ. J. Inorg. Chem.* **58**(13), 1605–1624 (2013).

4. Z. Ahmad, "The Properties and Application of Scandium-Reinforced Aluminum," *JOM* **55**(2), 35–39 (2003).

5. J. W. van Spronsen, *The Periodic System of the Chemical Elements: A History of the First Hundred Years*, Elsevier, Amsterdam, 260–284 (1969).

6. G. C. Pimentel and R. D. Sprately, *Understanding Chemistry*, Holden-Day, San Francisco, CA, 862 (1971).

7. T. Moeller, *The Chemistry of the Lanthanides: Pergamon Texts in Inorganic Chemistry* (Volume 26), Pergamon Press, New York, NY (1973).

8. C. J. Jones, *d- and f-block Chemistry: Tutorial Chemistry Texts*, Royal Society of Chemistry, London, England (2001).

9. H. C. Aspinall, *Chemistry of the f-Block Elements: Advanced Chemistry Texts*, Gordon & Breech, Amsterdam (2001).

10. C.-H. Huang (Ed.), *Rare Earth Coordination Chemistry: Fundamentals and Applications*, John Wiley, Singapore (2010).

11. B. D. Cullity and C. D. Graham, *Introduction to Magnetic Materials*, 2nd ed., Wiley, New York, NY (2009).

12. G. A. Molander, "Application of Lanthanide Reagents in Organic Synthesis," *Chem. Revs.* **92**, 29–68 (1992).

13. W. J. Evans, "Perspectives in Reductive Lanthanide Chemistry," *Coord. Chem. Revs.* **206–207**, 263–283 (2000).

14. J.-C. G. Bünzli et al., "New Opportunities for Lanthanide Luminescence," *J. Rare Earths* **25**(3), 257–274 (2007).

15. K. Binnemans, "Lanthanide-Based Luminescent Hybrid Materials," *Chem. Rev.* **109**, 4283–4374 (2009).

16. J.-C. G. Bünzli and S. V. Eliseeva, "Intriguing Aspects of Lanthanide Luminescence," *Chem. Sci.* **4**, 1939–1949 (2013).

17. A. Pol et al., "Rare Earth Metals Are Essential for Methanotrophic Life in Volcanic Mudpots," *Environ. Microbiol.* **16**(1), 255–264 (2014).

18. N. E. Holden and T. Coplen, "The Periodic Table of Elements," *Chem. Int.* **26**(1), 1 (2004).

19. E. D. Cater, "High Temperature Chemistry of Rare Earth Compounds: Dramatic Examples of Periodicity," *J. Chem. Educ.* **55**(11), 697–701 (1978).

20. D. A. Johnson, "Principles of Lanthanide Chemistry," *J. Chem. Educ.* **57**(7), 475–477 (1980).

21. M. Laing, "A Revised Periodic Table with the Lanthanides Repositioned," *Found. Chem.* **7**, 203–233 (2005).

22. P. F. Lang, "Is a Metal 'Ions in a Sea of Delocalized Electrons'?" *J. Chem. Educ.* **95**, 1787–1793 (2018).

23. D. A. Johnson and P. G. Nelson, "Valencies of the Lanthanides," *Found. Chem.* **20**, 15–27 (2018).

24. M. Seitz *et al.*, "The Lanthanide Contraction Revisited," *J. Am. Chem. Soc.* **129**, 11153–11160 (2007).

25. B. E. Douglas, "The Lanthanide Contraction," *J. Chem. Educ.* **31**(11), 598–599 (1954).

26. D. W. Pearce and P. W. Selwood, "Anomalous Valencies of the Rare Earths," *J. Chem. Educ.* **13**(5), 224–230 (1936).

27. D. R. Lloyd, "On the Lanthanide and 'Scandinide' Contractions," *J. Chem. Educ.* **63**(6), 502–503 (1986).

28. Q. Zhang *et al.*, "Pentavalent Lanthanide Compounds: Formation and Characterization of Praseodymium(V) Oxides," *Angew. Chem. Int. Ed. Engl.* **55**(24), 6896–6900 (2016).

29. S. Fabre *et al.*, "Paleoceanographic Significance of Cerium Anomalies during the OAE 2 on the NW African Margin," *J. Sediment. Res.* **88**(11), 1284–1299 (2018).

30. M. N. Bochkarev, "Molecular Compounds of 'New' Divalent Lanthanides," *Coord. Chem. Rev.* **248**(9–10), 835–851 (2004).

31. C. T. Palumbo *et al.*, "Molecular Complex of Tb in the +4 Oxidation State," *J. Am. Chem. Soc.* **141**(25), 9827–9831 (2019).

32. P. F. Lang and B. C. Smith, "Ionization Energies of the Lanthanides," *J. Chem. Educ.* **87**(8), 875–881 (2010).

33. R. Schmid, "The Noble Gas Configuration — Not the Driving Force but the Rule of the Game in Chemistry," *J. Chem. Educ.* **80**(8), 931–937 (2003).

34. P. Cann, "Ionization Energies, Parallel Spins, and the Stability of Half-Filled Shells," *J. Chem. Educ.* **77**(8), 1056–1061 (2000).

35. B. Johansson *et al.*, "Cerium: Crystal Structure and Position in the Periodic Table," *Sci. Rep.* **4**, 6398 (2014).

36. T. Ternström, "Subclassification of Lanthanides and Actinides," *J. Chem. Educ.* **53**(10), 629–667 (1976).

37. M. Laing, "Gadolinium: Central Metal of the Lanthanoids," *J. Chem. Educ.* **86**(2), 188–189 (2009).

38. O. Horovitz and C. Sârbu, "Characterization and Classification of Lanthanides by Multivariate-Analysis Methods," *J. Chem. Educ.* **82**(3), 473–483 (2005).

39. M. Laing, "Properties of the Lanthanide Metals: Correlations and Discontinuities," *J. Chem. Educ.* **82**(11), 1623 (2005).

40. K. C. Nicolaou *et al.*, "Samarium Diiodide-Mediated Reactions in Total Synthesis," *Angew. Chem. Int. Ed. Engl.* **48**(39), 7140–7165 (2009).

41. J. A. Cotruvo *et al.*, "Lanmodulin: A Highly Selective Lanthanide-Binding Protein from a Lanthanide-Utilizing Bacterium," *Biochemistry* **140**(44), 15056–15061 (2018).

42. M. Baum, "Rare-Earth Element Mobility during Hydrothermal and Metamorphic Fluid-Rock Interaction and the Significance of the Oxidation State of Europium," *Chem. Geol.* **93**(3–4), 219–230 (1991).

43. J. B. Thomas *et al.*, "Melt Inclusions in Zircon," *Rev. Mineral. Geochem.* **53**(1), 63–87 (2003).

Chapter 13

Actinoid and Post-Actinoid Elements

The location of the series of radioactive elements that we now call the actinoids was once a question in itself. These elements were first believed to be the commencement of a new d-block row, as a result of similarities to the chemistry of the corresponding elements above them. Then it was realized that they formed a new f-series of elements. However, it is sometimes overlooked that the earlier actinoids do indeed have strong resemblances to transition metals in their chemistry. This chapter also includes discussion of the post-actinoid elements as they fit better in this context.

In this chapter, the actinoids will be considered as encompassing the elements from 89 to 103. As even the longest lived isotopes of the later actinoids are highly radioactive, most of the discussions will be about the earlier actinoids that have very long-lived isotopes. In addition, the limited known chemistry of the post-actinoid elements will be contextualized.

The Actinoid Elements

Just as the lanthanoids were defined in Chapter 12, so it is necessary to define the actinoids. The commonly accepted definition of an actinoid is therefore:

> An **actinoid** is any of the series of 15 consecutive chemical elements in the Periodic Table from actinium to lawrencium (Figure 13.1).

Actinide Hypothesis

But first, a step back in time to consider the dispute on the location of the actinoid elements in the Periodic Table. Two of the heavy radioactive elements had been discovered by

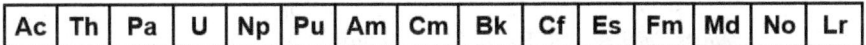

Figure 13.1 The actinoid elements as defined in this chapter.

Ряды.											
	\multicolumn{11}{c}{Г Р У П П Ы Э Л Е М Е Н Т О В Ъ:}										
	0	I	II	III	IV	V	VI	VII	\multicolumn{3}{c}{VIII}		
8	Ксе-нонъ. Xe 128	Це-зiй. Cs 132,9	Ба-рiй. Ba 137,4	Лан-танъ. La 138,9	Це-рiй. Ce 140,2	—	—	—	—	—	—
9	—	—	—	—	—	—	—	—	—	—	—
10	—	—	—	Иттер-бiй. Yb 173	—	Тан-талъ. Ta 183	Вольф-рамъ. W 184	—	Ос-мiй. Os 191	Ири-дiй. Jr 193	Пла-тина. Pt (Au) 194,8
11	Зо-лото. Au 197,2	Ртуть. Hg 200,0	Талiй. Tl 204,1	Сви-нецъ. Pb 206,9	Вис-мутъ. Bi 208,5	—	—				
12	—	—	Радiй. Rd 225	—	Торiй. Th 232,5	—	Уранъ. U 238,5				

Figure 13.2 The 8th through 12th Periods of Mendeléev's 1905 version of the Periodic Table.

the mid-19th century: uranium (1798) and thorium (1829). These featured as the sole members of the Series 10 in Mendeléev's 1871 Periodic Table, in Groups IV and VI respectively. By the 1905 version of his Periodic Table, radium (then Rd) had been discovered (1899). This element was placed in the Group II location in what had become Series 12 (Figure 13.2) [1].

By the early 20th century, two more elements were added to the series: actinium and protactinium, the "missing" elements exhibiting +3 and +5 oxidation-state chemistry. Cotton has commented [2]:

> So on its appearance in 1938, Emeléus and Anderson's "Modern Aspects of Inorganic Chemistry" (which was to become the leading inorganic chemistry textbook of the day) [3] printed a Periodic Table on page 2

which showed the four known actinides (although they did not refer to them as that) Ac, Th, Pa, and U in groups III-VI respectively.

It was in 1942 that Villar queried this assignment. He first made a general proposal for a revision of the 7th Period [4]:

If the sixth and seventh periods are made up of an equal number of elements, they should have identical configurations and therefore there must exist in the seventh period an array of 15 elements similar to the rare earths which, by analogy, should occupy the place reserved so far for actinium alone (Z = 89).

In a follow-up article, Villar focused upon the similarities of thorium to the rare earth elements, especially cerium. One potential problem that he posed was the dominance of the +4 oxidation state for thorium compared with +3 for the rare earths. He answered his own question [5]:

It is important to note that the element which occupies the second place in the rare earth series and which would be the homolog of thorium in the actinium series is cerium, which is characterized by being tri- and tetravalent. ... besides, cerium compounds in general have the same empirical formulas as the corresponding thorium compounds.

Unfortunately, Villar's contribution was totally overlooked, as was the *Actinide Hypothesis* championed in French journal articles by Janet in 1928 [6].

During the 1940s, four more elements were synthesized: neptunium, plutonium, americium, and curium. However, their respective chemistries did not correspond at all with those of the matching transition metal series. In fact, with the predominant oxidation state of +3, these, too, more resembled the lanthanoids. It was Seaborg who gained fame for pronouncing that the elements had been assigned the wrong location in the Periodic Table. In an article in *Science*, he displayed a version of the Periodic Table (Figure 13.3) with the post-radium elements located both as

1 H 1.008																1 H 1.008	2 He 4.003
3 Li 6.940	4 Be 9.02											5 B 10.82	6 C 12.010	7 N 14.008	8 O 16.0.C	9 F 19.00	10 Ne 20.181
11 Na 22.997	12 Mg 24.32											13 Al 26.97	14 Si 28.06	15 P 30.98	16 S 32.06	17 Cl 35.457	18 A 39.944
19 K 39.096	20 Ca 40.08	21 Sc 45.10	22 Ti 47.90	23 V 50.95	24 Cr 52.01	25 Mn 54.93	26 Fe 55.85	27 Co 58.94	28 Ni 58.69	29 Cu 63.57	30 Zn 65.38	31 Ga 69.72	32 Ge 72.60	33 As 74.91	34 Se 78.96	35 Br 79.916	36 Kr 83.7
37 Rb 85.48	38 Sr 87.63	39 Y 88.92	40 Zr 91.22	41 Cb 92.91	42 Mo 95.95	43	44 Ru 101.7	45 Rh 102.91	46 Pd 106.7	47 Ag 107.880	48 Cd 112.41	49 In 114.76	50 Sn 118.70	51 Sb 121.76	52 Te 127.61	53 I 126.92	54 Xe 131.3
55 Cs 132.91	56 Ba 137.36	57–71 La SEE	72 Hf 178.6	73 Ta 180.88	74 W 183.92	75 Re 186.31	76 Os 190.2	77 Ir 193.1	78 Pt 195.23	79 Au 197.2	80 Hg 200.61	81 Tl 204.39	82 Pb 207.21	83 Bi 209.00	84 Po	85	86 Rn
87	88 Ra	89 AC SEE	90 Th	91 Pa	92 U	93 Np	94 Pu	95 Am	96 Cm								

LANTHANIDE SERIES	57 La 138.92	58 Ce 140.13	59 Pr 140.92	60 Nd 144.27	61	62 Sm 150.43	63 Eu	64 Gd 156.9	65 Tb 159.2	66 Dy 162.46	67 Ho 163.5	68 Er 167.2	69 Tm 169.4	70 Yb 173.04	71 Lu 174.99

ACTINIDE SERIES	89 Ac	90 Th 232.12	91 Pa 231	92 U 238.07	93 Np 237	94 Pu	95 Am	96 Cm							

Figure 13.3 Seaborg's 1946 version of the Periodic Table (from Ref. [7]).

the traditional continuation of the d-series and as a new "actinide" set [7].

Seaborg commented [7]:

> ... I do want to say that the evidence strongly indicates that we are dealing here with a transition series of elements in which the 5f electron shell is being filled in a manner similar to the filling of the 4f electron shell in the well-known rare earth series. Apparently this new transition series begins with actinium in the same sense that the rare earth series begins with lanthanum, and, although the first elements in the heavy series exhibit the property of undergoing oxidation to higher oxidation states up to a maximum oxidation state of VI, the tendency in the later members is to have a stable lower oxidation state, such as the III state.

Laing showed that, during 1945, Sacks had drawn the same conclusion from studying trends, not in orbital occupancy, but in physical properties [8]:

> Uranium seemed like a transition metal, seemed like eka-tungsten—and yet, I felt somehow uncomfortable about this, and decided to do a little exploring, to examine the densities and melting points of all the transition

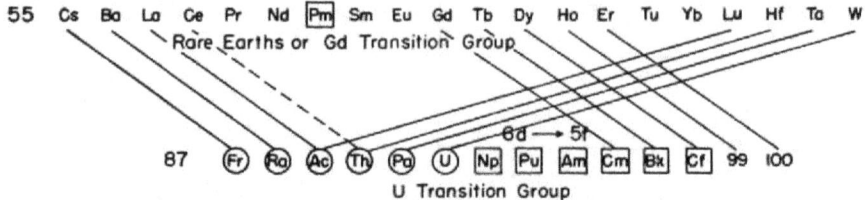

Figure 13.4 Part of Coryell's Periodic Table showing linkages of actinoids to earlier periods (adapted from Ref. [9]).

> metals. As soon as I did this I discovered an anomaly, for where the densities of the metals steadily increased through Periods 4, 5, and 6, they unexpectedly declined when one came to the elements in Period 7. Uranium was actually less dense than tungsten, though one would have expected it to have been more so (thorium, similarly, was less dense than hafnium, not more so, as one would have expected). It was precisely the same with their melting points, these reached a maximum in Period 6, then declined. … Could it be that these elements were instead the beginning of a second rare-earth series precisely analogous to the first one in Period 6?

The problem remained, as will be discussed in the following, that the early actinoids did indeed have high oxidation-state chemistry, which resembled that of the corresponding heavy transition metals. To recognize this fact, Coryell devised a Periodic Table, part of which is shown in Figure 13.4, which showed "tie-lines" to indicate the relevant linkages (the circled symbols are natural radioactive elements, while the boxed ones are synthetic).

Oxidation States of the Actinoids

The pattern of the highest common oxidation states of the early actinoids reflects the loss of all outer electrons, and this pattern parallels that of the transition metals more than that of the lanthanoids. For the later actinoids, +3 oxidation state is usually dominant, as with the lanthanoids. However, an additional oxidation state that results in an f^7 or f^{14} configuration can be found for some of the actinoids.

Kaltsoyannis and Kerridge have commented [10]:

> ... *the 5f orbitals of the early actinide elements have a larger radial extension than their 4f counterparts, and 5f involvement in the bonding of compounds of the first few actinides is not uncommon. Hence the bonding in compounds of, for example, uranium, is often found to be more covalent than in analogous compounds of the lanthanide series. ... It is generally accepted that as the actinide series is crossed, the chemistry increasingly resembles that of the lanthanides; ionic bonding and a single (trivalent) oxidation state dominate.*

The first five actinoids definitely have the highest oxidation state as corresponding to the noble gas electron configuration (see Table 13.1). Naturally, there has been much interest in whether the series can be completed with plutonium(VIII) compounds but as of the date of writing, no stable solid plutonium(VIII) compound had been isolated. There is evidence from spectroscopic measurements of solution species [11] and also from computational studies [12] that Pu(VIII) species should be isolable.

Table 13.1   Atom configurations and highest oxidation states for the early actinoids

Element	Atom Configuration	Charge on Noble Gas Configuration Ion
Actinium	[Rn]$7s^2 5f^0 6d^1$	+3
Thorium	[Rn]$7s^2 5f^0 6d^2$	+4
Protactinium	[Rn]$7s^2 5f^2 6d^1$	+5
Uranium	[Rn]$7s^2 5f^3 6d^1$	+6
Neptunium	[Rn]$7s^2 5f^4 6d^1$	+7
Plutonium	[Rn]$7s^2 5f^6$	+8(?)

Table 13.2 Atom configurations and highest oxidation states for elements 101–103

Element	Atom Configuration	Charge on [Rn]5f^{14} Configuration Ion
Mendelevium	[Rn]7s^{2}5f^{13}	+1
Nobelium	[Rn]7s^{2}5f^{14}	+2
Lawrencium	[Rn]7s^{2}5f^{14}7p^1	+3

Beyond, plutonium, americium has common oxidation states up to +7; curium, +4; berkelium, +4; while for the later actinoids, +3 predominates. There are two exceptions. First, the most stable oxidation state of nobelium is +2 [13]. Second, surprisingly, the +1 oxidation state is found for mendelevium [14]. It is stated that the +1 ionic radius is about 120 pm, in the range of the sodium and potassium ions. Both of these oxidation states can be accounted for by considering that they correspond to the full 5d^{14} set with the other outer orbitals empty (Table 13.2). For lawrencium, the 7p^1 electron has an extremely low ionization energy [15], yet its chemistry is dominated by the +3 oxidation state.

The Early Actinoid Relationships with Transition Metals

As discussed earlier in this chapter, there are significant resemblances of the early actinoids to transition metals. We can state this similarity as:

> The **actinoid relationship** *relates to similarities in chemical formulas and chemical properties between early members of the actinoid series and the corresponding members of the transition metal series.*

Hoffman and Lee reconstructed the Periodic Table to illustrate the closer links of the early actinoids, than lanthanoids with the corresponding transition metal group. They placed the actinoids above the lanthanoids. By stepping the early actinoids, Hoffman and Lee hoped to indicate the degree of resemblance [17]. These similarities diminished from thorium to plutonium, the only link from the lanthanoids to the transition metals being cerium (Figure 13.5).

To illustrate some of the transition metal/actinoid similarities, Table 13.3 shows comparative formulas of some compounds and polyatomic ions of uranium(VI) with those of the Group 6 metals in the +6 oxidation state. However, similarity in formula does not necessarily mean similarity in bonding and structure. For example, the molybdenyl cation, $[MoO_2]^{2+}$ has a bent geometry, while the uranyl cation $[UO_2]^{2+}$ is linear [16].

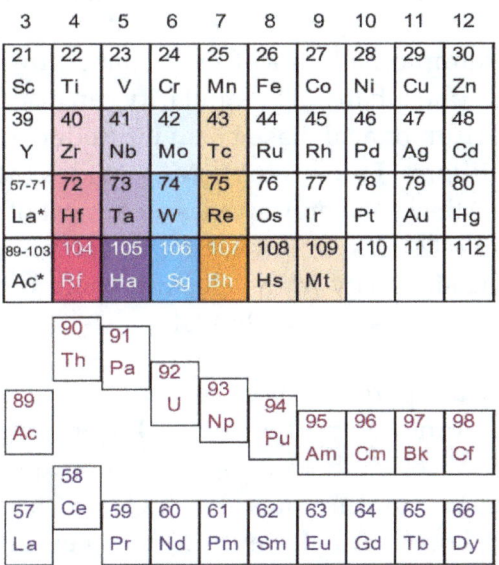

Figure 13.5 Relationship of early actinoid (and early lanthanoid) elements to the transition metals (adapted from Ref. [17]).

Table 13.3 A comparison of some species of uranium(VI) with those of the (VI) oxidation state of Group 6 members

Uranium Species	Group 6 Analogue
$[U_2O_7]^{2-}$ (yellow)	$[Cr_2O_7]^{2-}$ (orange)
UO_2Cl_2	CrO_2Cl_2, MoO_2Cl_2
UCl_6	WCl_6

Similarities of Thorium(IV), Cerium(IV), and Hafnium(IV)

Hoffman and Lee specifically noted (as had Villar, decades earlier) that not only did thorium(IV) chemistry strongly resemble that of zirconium(IV) and hafnium(IV) of the Group 4 elements, it also resembled that of cerium(IV) [17].

As examples, all four of these ions form insoluble fluorides and phosphates. The similarity can also be noted from a comparison of acid–base behavior under strongly oxidizing conditions. The insoluble MO_2 oxides (M = Th^{4+}, Ce^{4+}, Zr^{4+}, Hf^{4+}) being the only species over most of the pH range, the soluble M^{4+} ion only existing under very acidic conditions. The similarity in the electron configurations of their 4+ ions can be seen in Table 13.4.

There are particularly strong similarities in the chemistry of cerium(IV) and thorium(IV). Cerium(IV) oxide and thorium(IV) oxide both adopt the fluorite structure. They form isostructural nitrates, $M(NO_3)_4 \cdot 5H_2O$, where M is Ce or Th, and both form hexanitrato-complex ions $[M(NO_3)_6]^{2-}$. The major difference between the two elements in this oxidation state is that thorium(IV) is the thermodynamically stable form of that element while cerium(IV) is strongly

Table 13.4 The 4+ ions and their corresponding electron configurations

Ion	Noble Gas Core Electron Configuration
Zirconium(IV)	[Kr]
Cerium(IV)	[Xe]
Hafnium(IV)	[Xe]4f^{14}
Thorium(IV)	[Rn]

oxidizing (E^θ = +1.44 V). There is a mineralogical link between cerium(IV) and thorium(IV). Thorium and the lanthanoids — particularly cerium — are found together in two minerals, *monazite* and *xenotime* [18].

Similarities of the Later Actinoids with the Lanthanoids

There is a generic similarity of the later actinoids with the lanthanoids as they share the common oxidation state of +3. As an example of a close parallel, Xu and Pyykkö have shown that each of the first three ionization potentials of lawrencium is very close to those of the lanthanoid analogue, lutetium [19].

One specific correlation was demonstrated by Thompson *et al.* in the context of chromatographic elution curves (Figure 13.6). As can be seen there is a remarkable match in peaks for the corresponding pairs of terbium–berkelium, gadolinium–curium, and europium–americium [20].

According to computational studies, one difference between a lanthanoid(III) ion and the corresponding actinoid ion is the degree of bond covalency involving f orbitals. In this investigation, a comparison of the

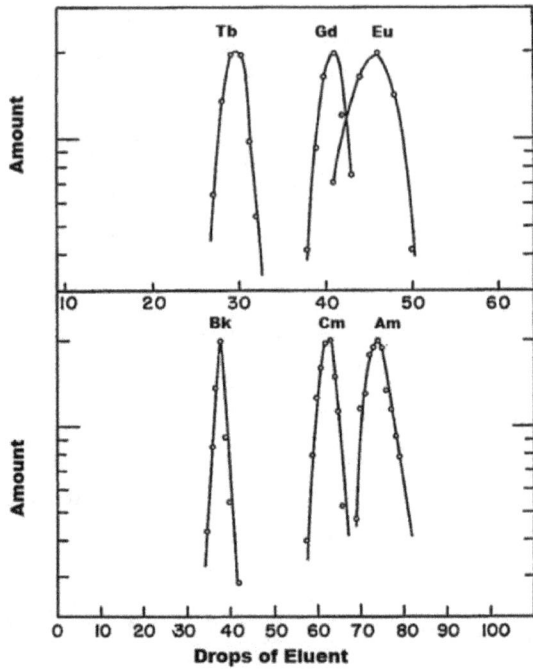

Figure 13.6 A comparison of chromatographic elution curves for three lanthanoids and three actinoids (adapted from Ref. [20]).

bonding in valence-isoelectronic $[EuCl_6]^{3-}$ and $[AmCl_6]^{3-}$ was made. Cross *et al.* concluded that the involvement of the americium 5f electrons in bonding with the 3d electrons of chlorine is far more significant than any involvement of the europium 4f electrons with the 3d orbitals of chlorine [21].

Similarity of Nobelium(II) and Group 2 Elements

As was mentioned earlier, nobelium favors the +2 oxidation state. This preference is far more than that of ytterbium, the corresponding lanthanoid, for which the +2 oxidation state is readily oxidized.

Table 13.5 Atom and 2+ ion configurations for radium and nobelium

Element	Atom Configuration	+2 Ion Configuration
Radium	$[Rn]7s^2$	$[Rn]$
Nobelium	$[Rn]7s^25f^{14}$	$[Rn]5f^{14}$

However, somewhat surprisingly, there is a much greater similarity to the chemistry of the lower alkaline earth metal ion as Maly *et al.* have stated [22]:

> In the absence of oxidizing or reducing agents the chromatographic and coprecipitation behavior of element 102 is similar to that of the alkaline earth elements. ... Nobelium is the first actinide for which the +2 oxidation state is the most stable species in aqueous solution.

The similarity in electron configuration with radium can be seen from Table 13.5.

Post-Actinoid Elements

The post-actinoid elements or, more correctly, the super-heavy elements (see Chapter 5), are those from element 104 (rutherfordium) up to yet-to-be-discovered element 126. From rutherfordium to copernicium, these are the 6d members of the transition metal series, then nihonium to oganesson correspond to the filling of the 7p orbitals.

With ever shorter half-lives leading to the ephemeral elements (see Chapter 5), the knowledge of their actual chemistry is very limited. Computational studies of what their chemistry should be increasingly fill the literature, but these unsubstantiated claims will be given only limited space here.

The 6d Elements

With isotope half-lives up to 1.3 hr, aspects of the chemistry of rutherfordium are well established. In particular, the +4 ion seems to be the sole common oxidation state, corresponding to an $[Rn]5f^{14}$ electron configuration. The compounds characterized to the date of writing this book match those of the heavier Group 4 elements (zirconium and hafnium), specifically: $RfCl_4$, $RfBr_4$, $RfOCl_2$, and K_2RfCl_6 [23].

Proceeding along the 6d series, the chemistry is based on every more limited data. Dubnium seems to behave like its Group 5 "relatives" niobium and tantalum, and also "pseudo-homologue" protactinium [24]. Likewise, seaborgium appears to form SgO_2Cl_2 analogous to MoO_2Cl_2 and WO_2Cl_2 [25]. Synthesis of $Sg(CO)_6$ matching with $Mo(CO)_6$ and $W(CO)_6$, has also been claimed [26]. Lougheed has commented that, despite predictions of relativistic effects causing dramatic changes in the chemical behavior of these elements, the chemistry of the 6d elements established so far seems to be that of their corresponding earlier group members. Seaborgium, in particular, from its limited chemistry, seems to be just a "normal" Group 6 element [27].

Bohrium, too, follows the pattern of having a matching chemical compound to the other heavy transition metals of Group 7. That is, it forms the compound BhO_3Cl, analogous to TcO_3Cl and ReO_3Cl [28]. Similarly, hassium has been shown to form the characteristic species of Group 8, that is, HsO_4 [29].

With such short-lived isotopes, meitnerium seems to mark the current limit of practical chemistry. The chemistry of this element would be of particular interest. Assuming that it did behave chemically as a member of Group 9, it has been proposed from theoretical studies that it might form a compound of meitnerium(IX), that is, $[MtO_4]^+$, isoelectronic to HsO_4 [30].

A theoretical study of the properties of darmstadtium indicated that it should resemble platinum of Group 10 in its chemistry, in particular in forming a strong bond to the carbonyl ligand [31]. Similarly, roentgenium is predicted to resemble silver of Group 11 in readily forming a +1 species in the $[Rg(OH_2)_2]^+$ ion [32]. Likewise, according to computational analysis, copernicium is likely to have physical and chemical properties resembling those of mercury [33].

The 7p Elements

All of the presumptions of the chemistry of these elements come from theoretical computations. As such, and as some are contradictory, only a few selected cases will be described. Nihonium is expected to be a typical Group 13 element with a predominant +3 oxidation state [34] while it has been proposed that flerovium is a volatile metal [35]. The most attention has focused on oganesson, the Group 18 member of the period. It has been suggested that this element would be a liquid at room temperature [36]. It has also been proposed that cooled to its solid phase, oganesson would be a semiconductor [37].

And Beyond . . .

With the synthesis of element 117, the 7th Period has been completed. In Chapter 1, the issue of the future was addressed from the nucleosynthesis aspect. From the chemistry perspective, it is not about synthesizing a few fleeting atoms, but about finding long-lived isotopes — if they exist.

Is there an "island of stability" yet to be found? If so, can super-neutron-rich projectiles be synthesized to bombard superheavy nuclei and reach the "island" [38]? Though such claims have been made before, it really does seem that the limit of actual chemistry of any new element has been reached. Yet there is the tantalization that such elements would open new possibilities in electron structure. Beyond the 8s orbitals loom the possibility of the 5g set. What are the properties of elements with g electrons? Will the orbitals fill in sequence or will things become "messy" as Pyykkö has suggested, in what is now called the *Pyykkö Model* [39].

Gilead has commented upon the "eka-elements" — those which are not actually known [40]:

> There is no guarantee that they [eka-elements] will eventually be discovered, synthesized, or isolated as actual.

Perhaps this is indeed the completion of elemental chemistry as chemists know it.

Commentary

The actinoids exhibit a wider variety of chemical behavior than do the lanthanoids. Even though the first members of the actinoid series are no longer placed with the transition metals, nevertheless, in the highest oxidation states, there are strong chemical similarities to them. In addition, toward the end of the series, the preference for a $5f^{14}$ electron configuration marks results in two surprisingly low stable oxidation states. Unfortunately, the short half-lives of these later actinoids decrease the ability to fully explore their chemistry. Despite the strong influence of relativistic effects, even for the post-actinoid elements, the element properties so far seem to be mostly those expected for the appropriate group membership.

Will this continue into the 8th Period? As Haba has commented [41]:

> *Owing to the predicted strong influence of relativistic effect, any experimental investigation of their properties is fascinating.*

Is attaining stable enough atoms of elements of the 8th Period a feasible goal or simply fantasy? Only time will tell.

References

1. P. J. Stewart, "Mendeleev's Predictions: Success and Failure," *Found. Chem.* **21**, 3–9 (2019).
2. S. A. Cotton, Personal Communication, 06 October 2019.
3. H. J. Emeléus and J. S. Anderson, *Modern Aspects of Inorganic Chemistry*, Routledge & Sons Ltd., London, England (1938).
4. G. E. Villar, "On a Suggested Revision of the Seventh Period of the Periodic Table," *J. Chem. Educ.* **19**(6), 286 (1942).
5. G. E. Villar, "A Suggested Revision of the Position of Thorium in the Fourth Period of the Periodic Table," *J. Chem. Educ.* **19**(7), 329–330 (1942).
6. P. J. Stewart, "Charles Janet: Unrecognized Genius of the Periodic System," *Found. Chem.* **12**, 5–15 (2010).
7. G. T. Seaborg, "The Transuranium Elements," *Science* **104**, 379–386 (1946).
8. M. J. Laing, "The Question Mark at Uranium," *Found. Chem.* **12**, 27–30 (2010).
9. C. D. Coryell, "The Periodic Table: The $6d$-$5f$ Mixed Transition Group," *J. Chem. Educ.* **29**, 62–64 (1952).
10. N. Kaltsoyannis and A. Kerridge, "11: Chemical Bonding of Lanthanides and Actinides," in G. Frenking and S. Shaik (Eds.), *The Chemical Bond: Chemical Bonding Across the Periodic Table*, Wiley-VCH, Weinheim, Germany, 342–343 (2014).
11. Yu. M. Kiselev *et al.*, "On Existence and Properties of Plutonium (VIII) Derivatives," *Radiochim. Acta* **102**(3), 227–237 (2014).
12. W. Huang *et al.*, "Is Octavalent Pu(VIII) Possible? Mapping the Plutonium Oxyfluoride Series PuO_nF_{8-2n} (n = 0–4)," *Inorg. Chem.* **54**(17), 8825–8831 (2015).

13. R. J. Silva *et al.*, "Comparative Solution Chemistry, Ionic Radius, and Single Ion Hydration Energy of Nobelium," *Inorg. Chem.* **13**(9), 2233–2237 (1974).

14. N. B. Mikheev, "Lower Oxidation States of Actinides," *Radiochim. Acta* **32**, 69–80 (1983).

15. A. Türler, "Lawrencium Bridges a Knowledge Gap," *Nature* **520**, 166–167 (2015).

16. K. Tatsumi and R. Hoffmann, "Bent cis d^0 MoO_2^{2+} vs. Linear trans d^0f^0 UO_2^{2+}: A Significant Role for Nonvalence 6p Orbitals in Uranyl," *Inorg. Chem.* **19**(9), 2656–2658 (1980).

17. D. C. Hoffman and D. M. Lee, "Chemistry of the Heaviest Elements — One Atom at a Time," *J. Chem. Educ.* **76**(3), 332–347 (1999).

18. H. E. Kremers, "Technology of the Rare Earths," *J. Chem. Educ.* **62**(8), 665–667 (1985).

19. W.-H. Xu and P. Pyykkö, "Is the Chemistry of Lawrencium Peculiar?" *Phys. Chem. Chem. Phys.* **18**, 17351–17355 (2016).

20. S. G. Thompson, A. Ghiorso, and G. T. Seaborg, "The New Element Berkelium (Atomic Number 97)," *Phys. Revs.* **80**, 781–789 (1950).

21. J. N. Cross *et al.*, "Covalency in Americium(III) Hexachloride," *J. Am. Chem. Soc.* **139**, 8667–8677 (2017).

22. J. Maly *et al.*, "Nobelium: Tracer Chemistry of the Divalent and Trivalent Ions," *Science* **160**, 1114–1115 (1968).

23. B. Kadkhodayan *et al.*, "On-Line Gas Chromatographic Studies of Chlorides of Rutherfordium and Homologs Zr and Hf," *Radiochim. Acta* **72**(4), 169–178 (1996).

24. L. Öhrström, "Brief Encounters with Dubnium," *Nat. Chem.* **8**, 986 (2016).

25. M. Schädel *et al.*, "Chemical Properties of Element 106 (Seaborgium)," *Nature* **388**, 55–57 (1997).

26. J. Even *et al.*, "Synthesis and Detection of a Seaborgium Carbonyl Complex," *Science* **345**(6203), 1491–1493 (2014).

27. R. Lougheed, "Oddly Ordinary Seaborgium," *Nature* **388**, 21–22 (1997).

28. R. Eichler *et al.*, "Chemical Characterization of Bohrium (Element 107)," *Nature* **407**, 63–65 (2000).

29. C. E. Düllman *et al.*, "On the Stability and Volatility of Group 8 Tetroxides, MO_4 (M = Ruthenium, Osmium, and Hassium (Z = 108))," *J. Phys. Chem. B* **106**(26), 6679–6684 (2002).

30. D. Himmel *et al.*, "How Far Can We Go? Quantum-Mechanical Investigations of Oxidation State +IX," *ChemPhysChem.* **11**(4), 865–869 (2010).

31. M. Patzschke and P. Pyykkö, "Darmstadtium Carbonyl and Carbide Resemble Platinum Carbonyl and Carbide," *Chem. Commun.* **17**, 1982–1983 (2004).

32. R. D. Hancock, L. J. Bartolotti, and N. Kaltsoyannis, "Density Functional Theory-Based Prediction of Some Aqueous-Phase Chemistry of Element 111. Roentgenium(I) Is the 'Softest' Metal Ion," *Inorg. Chem.* **45**(26), 10780–10785 (2006).

33. R. Eichler *et al.*, "Thermochemical and Physical Properties of Element 112," *Angew. Chem. Int. Ed. Engl.* **47**(17), 3262–3266 (2008).

34. M. Seth and P. Schwerdtfeger, "The Chemistry of Superheavy Elements. III. Theoretical Studies on Element 113 Compounds," *J. Chem. Phys.* **111**, 6422–6433 (1999).

35. A. Yakushev *et al.*, "Superheavy Element Flerovium (Element 114) Is a Volatile Metal," *Inorg. Chem.* **53**(3), 1624–1629 (2014).

36. R. M. MacRae and T. J. Kemp, "Oganesson: A Most Unusual 'Inert Gas'," *Sci. Prog.* **101**(2), 101–120 (2018).

37. J.-M. Mewes *et al.*, "Oganesson Is a Semiconductor: On the Relativistic Band-Gap Narrowing in the Heaviest Noble-Gas Solids," *Angew. Chem. Int. Ed. Engl.* **58**(40), 14260–14264 (2019).

38. J. Reedijk, "Row 7 of the Periodic Table Complete: Can We Expect More New Elements; and If So, When?" *Polyhedron* **141**, 1–4 (2018).

39. P. Pyykkö, "A Suggested Periodic Table up to $Z \leq 172$, Based on Dirac-Fock Calculations on Atoms and Ions," *Phys. Chem. Chem. Phys.* **13**, 161–168 (2011).

40. A. Gilead, "Eka-Elements as Chemical Pure Possibilities," *Found. Chem.* **18**, 183–194 (2016).

41. H. Haba, "A New Period in Superheavy-Element Hunting," *Nat. Chem.* **11**, 10–13 (2019).

Chapter 14

Pseudo-Elements

Most chemists regard the ammonium ion as behaving much like an alkali metal cation. Likewise, cyanide ion shows similarities to halide ions. In this chapter, the similarities and differences of these polyatomic ions to their group analogues will be explored. In addition, some less-common "pseudo-element" ions will be introduced.

… and so, Gentle Reader, the end of this particular voyage has been reached. Rest assured, Gentle Reader, there are other voyages out there to cross hitherto uncharted seas of patterns and trends in the Periodic Table. The Periodic Table is not, and never has been "set in stone." The Author has endeavored to show that, contrary to public belief and to the passing references to the Periodic Table in chemistry textbooks, it is a living evolving organism. The White Queen in *Alice Through the Looking Glass* had it correct (Figure 14.1).

Yes, Alice, there are indeed fantastical — and once thought impossible — aspects to the Periodic Table. For example, until

Alice laughed. 'There's no use trying,' she said: 'one can't believe impossible things.'

'I daresay you haven't had much practice,' said the Queen. 'When I was your age, I always did it for half-an-hour a day. Why, sometimes I've believed as many as six impossible things before breakfast.'

Figure 14.1 Alice and the White Queen, from *Alice Through the Looking Glass* [1].

1962, it was **known** that it was **impossible** to make any compound of the noble gas elements. There was the "**proof**" of the octet-rule limit — so how could they? In some chapters, it has been proposed that elements belonged in more than one place in the Periodic Table — or in a place other than their atomic number/electron configuration mandated location — heresy in the past. The awesomeness of the Periodic Table continues in this final chapter: compounds and polyatomic ions that "behave" like elements and element ions.

Pseudo-Elements

Some polyatomic ions resemble element ions in their behavior, and, in a few cases, there is a molecule that corresponds to the matching element. We can define this unusual category as:

*A **pseudo-element** is the parent of a polyatomic ion whose behavior in many ways mimics that of an ion of an element or of a group of elements.*

In this chapter, the focus will be on the ammonium ion as a pseudo-alkali metal ion, and on the cyanide ion as a pseudo-halide ion (Figure 14.2).

The Ammonium Ion as a Pseudo-Alkali Metal Ion

Even though the ammonium ion is a polyatomic cation containing two nonmetals, it behaves in many respects like an alkali metal ion. The similarity results from the ammonium ion being a large low-charge cation just like the cations of the alkali metals. In fact, the radius of the ammonium ion (151 pm) is very close to that of the potassium ion (152 pm). However, the chemistry of ammonium salts more resembles that of rubidium or cesium ions, perhaps because the ammonium ion is not spherical, and

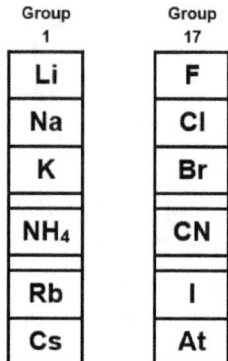

Figure 14.2 Relationship of ammonium to the Group 1 elements and of cyanide to the Group 17 elements.

its realistic radius is larger than its measured value. The similarity to the heavier alkali metals is particularly true of the crystal structures. Ammonium chloride, like rubidium chloride and cesium chloride, has a CsCl crystal lattice at high temperatures and a NaCl crystal lattice at low temperatures.

The ammonium ion also resembles an alkali metal ion in its precipitation reactions. Although all simple sodium compounds are water soluble, there are insoluble compounds of the heavier alkali metal ions with very large anions. The ammonium ion gives precipitates with solutions of these same anions. A good example is the hexanitritocobaltate(III) ion, $[Co(NO_2)_6]^{3-}$, which is commonly used as a test in qualitative analysis for the heavier alkali metals. With ammonium ion, a bright yellow precipitate of $(NH_4)_3[Co(NO_2)_6]$ is obtained analogous to that of $K_3[Co(NO_2)_6]$ with potassium ion.

However, the similarity does not extend to all chemical reactions that these ions undergo. For example, gentle heating of alkali metal nitrates typically gives the corresponding nitrite and oxygen gas, but heating ammonium nitrate results in decomposition of the cation and anion to give dinitrogen oxide and water.

Other Cations as Pseudo-Alkali Metal Ions

To stabilize large mononegative ions in the solid phase, a larger monopositive cation is required. One of the commonly used cations is the tetramethylammonium ion, $[N(CH_3)_4]^+$. With an ionic radius of 234 pm, this ion is even larger than the cesium ion (181 pm). As a result, the tetramethylammonium ion will form a stable solid compound with the very large pentaiodide(1−) ion, I_5^- [2], while the cesium ion will only form a solid cesium triiodide(1−), CsI_3.

In addition to the ammonium ion, the phosphonium ion, PH_4^+, had been used as a large monopositive cation [3]. Now the methyl-substituted phosphorus (and arsenic) analogues have become more popular (and safer) for this function [4]. With the very big tetraphenylphosphonium ion, $[P(C_6H_5)_4]^+$, the extremely large heptaiodide(1−) ion, I_7^- ion can be stabilized, forming $[P(C_6H_5)_4]I_7$.

Ammonium as a Species

A major weakness of the parallel between ammonium ion and the heavier alkali metal ions is that the parent pseudo-element of the ammonium ion, "NH_4," cannot be isolated. However, claims of the existence of "ammonium" in the form of a mercury amalgam date back to the early 1800s. The discovery of this species was claimed first by Davy, while experiments with ammonium amalgam were described by Wetherill in 1865 [5]. An electrolysis experiment using mercury and ammonium ion published in 1929 concluded that the only explanation for the cell reaction was [6]:

> ... the ammonium group can reach the cathode only by diffusing through the mercury in the form of a dissolved metal, since ammonia and ammonium salts are quite insoluble in mercury.

A reported electrolytic synthesis in 1951 [7], together with an article titled "Is There a Neutral Ammonium Radical?" in 1968 [8] marked a revival of interest in the possibility of such a species. Whiteside, Xantheas, and Gutowski showed from computational studies that the ammonium radical should resemble sodium in terms of its electronegativity [9]. The most convincing synthetic evidence of a pseudo-alkali metal radical has been that of tetramethylammonium with mercury. This species has been shown to have the formula $[N(CH_3)_4]Hg_8$ [10].

The explanation of the ammonium amalgam was reported in 1986 by Garcia *et al.* [11]. They provided convincing evidence that for the tetrabutylammonium amalgam, the structure was actually: $([N(t\text{-Butyl})_4]^+[Hg_4]^-$. By comparison, the structure of the tetramethylammonium amalgam should be: $[N(CH_3)_4]^+[Hg_8]^-$. Thus, the claimed ammonium radical does not exist but instead, the mercury cluster is another case of a metal with a significant electron affinity (see Chapter 2). In fact, the stability of mononegative mercury clusters has been investigated and their stability confirmed [12]. In fact, the clustering up to eight atoms results in a considerably increased (more negative) electron affinity, explaining why these anionic metal clusters so readily form [13].

Pseudo-Halogens

As ammonium and its relatives are to alkali metal cations, so pseudo-halogens (a term devised in 1925) are to halide anions. The following definition is appropriate:

> A **pseudo-halogen** is a polyatomic analogue of the Group 17 elements whose chemistry resembles that of one or more of the halogens. The polyatomic **pseudo-halide ion** may substitute for a halide ion and the resulting compounds should resemble in their chemistry those of the equivalent halide compound.

Denoting a pseudo-halide ion as (PseudoX)⁻ and a halide ion as X⁻, corresponding pseudo-halogen molecules of the forms (PseudoX)–(PseudoX) and (PseudoX)–X can form. As an example, the pseudo-halide ion, C≡N⁻, can be oxidized to form its own diatomic "parent" molecule, N≡C–C≡N, or "partner" with a halogen, such as chlorine, to form N≡C–Cl.

Among the more common pseudo-halide ions are the valence-isoelectronic series of cyanate, OCN⁻; thiocyanate, SCN⁻, and selenocyanate, SeCN⁻. The azide ion, N_3^-, is also considered a pseudo-halide ion even though its parent pseudo-halogen does not exist.

Those listed earlier are five of the "traditional" pseudo-halide ions. There is continuing interest in this category of ions/compounds [14]. One of the newer additions is the $(CS_2N_3)^-$ ion. This ion satisfies the full criteria in that the parent pseudo-halogen, $(CS_2N_3)_2$ has been synthesized as has an inter-pseudo-halogen, $(CS_2N_3–CN)$ [15].

Cyanide Ion as a Pseudo-Halide Ion

Earlier, it was described how the ammonium ion, despite being a polyatomic ion, behaved much like an alkali metal ion. However, the best example of a pseudo-element ion is cyanide. Not only does it behave very much like a halide ion but also the parent pseudo-halogen, cyanogen, $(CN)_2$, exists.

The cyanide ion resembles a halide ion in a remarkable number of ways:

- Compounds of cyanide ion with silver, lead(II), and mercury(I) ions are insoluble, as are those of chloride, bromide, and iodide ions.
- Just as solid silver chloride reacts with ammonia to give the diamminesilver(I) cation, so does silver cyanide.

- The cyanide ion is the conjugate base of the weak acid hydrocyanic acid, HCN, parallel to fluoride ion being the conjugate base of the weak acid, hydrofluoric acid.

The existence of cyanogen, C_2N_2, the "parent element" makes a stronger case for the concept of pseudo-elements.

- Cyanide ion can be oxidized to cyanogen in a similar manner to the oxidation of halides to halogens. The parallel is particularly close with iodide ion since they can both be oxidized by very weak oxidizing agents such as the copper(II) ion.
- Cyanogen reacts with base to give the cyanide ion and cyanate ion (CNO^-), in a parallel manner to the reaction of dichlorine with base to give chloride and hypochlorite ion (ClO^-).

Cyanogen forms pseudo-interhalogen compounds such as iodine monocyanide, ICN, in the same way that halogens form interhalogen compounds such as iodine monochloride, ICl.

The Tetracarbonylcobaltate(−I) Ion

This pseudo-halide ion was first synthesized in the 1930s. It was shown that the compound tetracarbonylhydrocobalt(−I), $HCo(CO)_4$, is highly acidic, with a pK_a of 8.5 [16]. In fact, its synthesis is by acidification of sodium tetracarbonylcobaltate(−I). As a result of its acidity, it was the first catalyst employed for the hydroformylation of alkenes (the oxo reaction) [17].

A Cautionary Note

To end with a cautionary note: terminology. Before claiming a pseudo-element is a "pseudo-alkali metal" or a "pseudo-halogen," it should meet specific criteria. Does it indeed have many matching chemical properties of one of those elemental groups; or is it simply a large polyatomic monopositive cation or mononegative anion?

Combo Elements

The compound carbon monoxide has several similarities to dinitrogen, N_2. For example, they are both triply bonded molecules with similar boiling points: $-196°C$ (N_2) and $-190°C$ (CO). A major reason for the parallel behavior is that the dinitrogen molecule and the carbon monoxide molecule are isoelectronic. This similarity extends to the chemistry of the two molecules. In particular, there are several transition metal compounds where dinitrogen can substitute for a carbon monoxide entity. For example, it is possible to replace one or two carbon monoxides bonded to chromium in $Cr(CO)_6$ to give isoelectronic $Cr(CO)_5(N_2)$ and $Cr(CO)_4(N_2)_2$.

The *combo elements* are a subset of isoelectronic behavior in which the sum of the valence electrons of a pair of atoms of one element (above being nitrogen) matches the sum of the valence electrons of two horizontal neighboring elements (carbon and oxygen):

*A **combo element** can be defined as the combination of an $(n - x)$ group element with an $(n + x)$ group element to form compounds that parallel those of the (n) group element.*

Boron–Nitrogen Analogs of Carbon Compounds

The best example of a combo element is that of the boron and nitrogen combination matching with a pair of carbon atoms, boron having one less valence electron than carbon, and nitrogen one more.

The simplest analogue is that of boron nitride, BN, for the graphite allotrope of carbon. Unlike graphite, boron nitride is a white solid that does not conduct electricity. This distinction is probably a result of the different way the layers in the two crystals are stacked. The layers in the graphite-like form of boron nitride are an almost identical distance apart as those in graphite. However, the boron nitride layers are organized so that the nitrogen atoms in one layer are situated directly over boron atoms in the layers above and below, and *vice versa*. This arrangement is logical, because the partially positive boron atoms and partially negative nitrogen atoms are likely to be electrostatically attracted to each other. By contrast, the carbon atoms in one layer of graphite are directly over the center of the carbon rings in the layers above and below.

In a further analogy to carbon, application of high pressures and high temperatures converts the graphite-like allotrope of boron nitride to diamond-like forms. The polymorph [18] with a zinc-blende type structure was first synthesized in 1957 [19] while the wurtzite form was not prepared until 2009 [20]. Thus, there are two allotropes, one with two polymorphs of boron nitride to match with those of carbon.

There has been an increasing interest in hybrid boron–nitrogen/carbon materials to benefit from the properties of both. An example is the coating of carbon nanotubes with a layer of boron nitride [21].

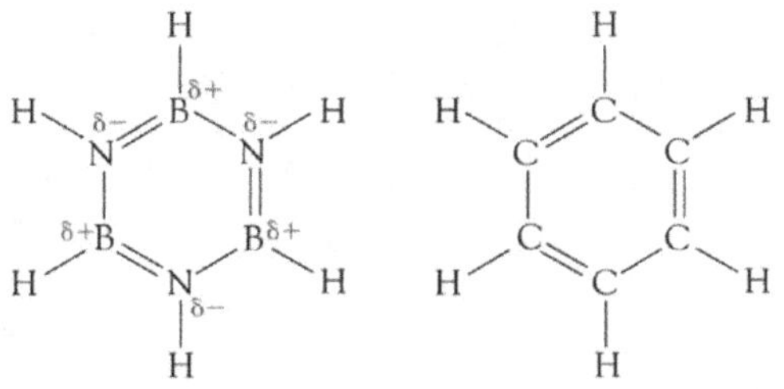

Borazine   Benzene

Figure 14.3   A comparison of benzene and borazine.

One of the other similarities between boron–nitrogen and carbon compounds is the compound borazine [22]. Borazine, $B_3N_3H_6$, is a cyclic molecule analogous to benzene, C_6H_6.

In fact, borazine is sometimes called "inorganic benzene" (Figure 14.3). This compound is a useful reagent for synthesizing other boron–nitrogen analogs of carbon compounds, but currently it has no commercial applications.

The polarity of the boron–nitrogen bond means that borazine exhibits localized π-bonding between pairs of atoms rather than the complete delocalized aromatic ring π-system of benzene [23]. Hence, borazine is much more prone to chemical attack than is the homogeneous ring of carbon atoms in benzene. Though the liquids have similar densities, there are significant differences in melting and boiling points (Table 14.1).

There has been interest in preparing other compounds in which the boron–nitrogen pair have been substituted for a carbon–carbon pair. For example, B–N substitutions have been made in diphenylacetylene [24]. These combo element replacements show that the polarity of the B–N bond influences the molecular structure. There is now increasing

Table 14.1 A comparison of some physical properties of benzene and borazine

	M. Pt. (°C)	B. Pt. (°C)	Density (g·cm⁻³)
Benzene	+5.5	+80	0.88
Borazine	−58	+53	0.81

interest in inorganic polymers employing a boron–nitrogen backbone [25].

Superatoms

Certain clusters of atoms can behave as if they were single entities. That is, free electrons in the cluster occupy a unique set of molecular orbitals. These bonding orbitals are all occupied, resulting in a "closed shell" system. Loss of one electron provides alkali metal ion behavior while addition of one electron results in halide ion behavior. The most comprehensive definition is:

> A **superatom** *is any cluster of atoms that seems to exhibit some of the properties of elemental atoms.*

The best documented superatoms are clusters of aluminum atoms generated with a negative charge. For example, it is claimed that the Al_{13} cluster behaves as a *super-halogen* [26]. Correspondingly, the $[Al_{13}I]^-$ cluster behaves like an inter-polyhalide ion, such as BrI^-. By contrast, the Al_{14}^{2+} cluster behaves as an alkaline earth metal.

Computational studies on the iron–oxygen cluster, FeO_4, have suggested that it, too, has a closed-shell structure. According to the computations, despite its closed-shell structure, this cluster has an electron affinity that is larger than that of any known halogen atom [27].

Synthetic Metals

Finally, a term, "synthetic metals" needs to be included here in this compilation for completeness, which, despite its name, does not really fit. The definition commonly accepted is [28]:

> A **synthetic metal** *possesses metallic conduction but is formed entirely of such nonmetallic atoms as carbon, nitrogen, hydrogen, the halogens and oxygen.*

The first use of this term was to describe ammonium amalgam (see earlier). However, the prototypical example of a synthetic metal is polysulfur nitride [29].

Commentary

Is this THE END? No, our perspectives on the patterns and trends in the Periodic Table will never become fixed. Until 1999, there was no knowledge of a knight's move relationship. Topological studies continue to find new unexpected linkages [30]. And the superatom clusters that have been synthesized, such as $[EuSn_6Bi_8]^{4-}$ [31], continue to push the boundaries beyond what a classical inorganic chemist would ever have believed possible. Maybe long-lived isotopes of ephemeral elements will be synthesized, and their actual chemistry studied. Perhaps, this book marks the END OF THE BEGINNING and that some of the thoughts therein will lead to the opening of new vistas for the future.

References

1. L. Carroll, *More Annotated Alice: Alice's Adventures in Wonderland and Through the Looking Glass and What Alice Found There*, with notes by Martin Gardner, Random House, New York, NY, 237 (1990).
2. C. A. L. Filgueiras *et al.*, "Tetramethylammonium Pentaiodide," *Acta Cryst.* **57**, 338–340 (2001).
3. H. G. Heal, "Analogues of the Ammonium Compounds in Periods Five, Six, and Seven," *J. Chem. Educ.* **35**(4), 192–197 (1958).
4. P. D. C. Dietzel, R. K. Kremer, and M. Jansen, "Superoxide Compounds of the Large Pseudo-Alkali-Metal Ions Tetramethylammonium, -Phosphonium, and -Arsonium," *Chem. Asian J.* **2**(1), 66–73 (2007).
5. C. M. Wetherill, "Experiments with the Ammonium Amalgam," *Chemical News,* 207–210 (3 November 1865).
6. J. H. Reedy, "Lecture Demonstration of Ammonium Amalgam," *J. Chem. Educ.* **6**(10), 1767 (1929).
7. R. J. Johnston and A. R. Ubbelohde, "The Formation of Ammonium by Electrolysis," *J. Chem. Soc.* 1731–1736 (1951).
8. J. K. S. Wan, "Is There a Neutral Ammonium Radical?" *J. Chem. Educ.* **45**(1), 40–43.
9. A. Whiteside, S. S. Xantheas, and M. Gutowski, "Is Electronegativity a Useful Descriptor for the Pseudo-Alkali Metal NH_4?" *Chem. Eur. J.* **17**, 13197–13205 (2011).
10. C. Hoch and A. Simon, "Tetramethylammonium Amalgam, $[N(CH_3)_4]Hg_8$," *Z. Anorg. Algem. Chem.* **632**, 2288–2294 (2006).
11. E. Garcia, A. H. Cowley, and A. J. Bard, "'Quaternary Ammonium Amalgams' as Zintl Ion Salts and Their Use in the Synthesis of Novel Quaternary Ammonium Salts," *J. Am. Chem. Soc.* **108**, 6082–6083 (1986).
12. Y. Wang, H.-J. Flad, and M. Dolg, "Structural Changes Induced by an Excess Electron in Small Mercury Clusters," *Int. J. Mass Spectrom.* **201**, 197–204 (2000).
13. J. Kang *et al.*, "Molecular Structures, Energetics, and Electronic Properties of Neutral and Charged Hg(n) Clusters (n = 2–8)," *J. Phys. Chem. A* **114**(18), 5630–5639 (2010).

14. H. Brand *et al.*, "Modern Aspects of Pseudohalogen Chemistry: News from CN– and PN–Chemistry," *Z. Anorg. Algem. Chem.* **633**(1), 22–35 (2007).

15. M.-J. Crawford *et al.*, "CS_2N_3. A Novel Pseudohalogen," *J. Am. Chem. Soc.* **122**(37), 9052–9053 (2000).

16. E. J. Moore, J. M. Sullivan, and J. R. Norton, "Kinetic and Thermodynamic Acidity of Hydrido Transition-Metal Complexes. 3. Thermodynamic Acidity of Common Mononuclear Carbonyl Hydrides," *J. Am. Chem. Soc.* **108**(9), 2257–2263 (1986).

17. I. Wender, H. W. Sternberg, and M. Orchin, "Evidence for Cobalt Hydrocarbonyl as the Hydroformylation Catalyst," *J. Am. Chem. Soc.* **75**, 3041–3042 (1953).

18. B. D. Sharma, "Allotropes and Polymorphs," *J. Chem. Educ.* **64**(5), 404–407 (1987).

19. R. H. Wentorf, "Cubic Form of Boron Nitride," *J. Chem. Phys.* **26**, 956 (1957).

20. Z. Pan *et al.*, "Harder than Diamond: Superior Indentation Strength of Wurtzite BN and Lonsdaleite," *Phys. Rev. Lett.* **102**(5), 055503 (2009).

21. Y. Wei *et al.*, "Polyimide Nanocomposites with Boron Nitride-Coated Multi-Walled Carbon Nanotubes for Enhanced Thermal Conductivity and Electrical Insulation," *J. Mat. Chem. A* **2**, 20958–20965 (2014).

22. D. A. Payne and E. A. Eads, "Boron-Nitrogen Heterocycles," *J. Chem. Educ.* **41**(6), 334–336 (1964).

23. R. Islas *et al.*, "Borazine: To Be or Not to Be Aromatic," *Struct. Chem.* **18**(6), 833–839 (2007).

24. A. J. V. Marwitz *et al.*, "BN-Substituted Diphenylacetylene: A Basic Model for Conjugated π-Systems Containing the BN Bond Pair," *Chem. Sci.* **3**, 825–829 (2012).

25. D. Resendiz Lara *et al.*, "Boron–Nitrogen Main Chain Analogues of Polystyrene: Poly(*B*-aryl)aminoboranes via Catalytic Dehydrocoupling," *Chem. Commun.* **53**, 11701–11704 (2017).

26. D. E. Bergeron, "Al Cluster Superatoms as Halogens in Polyhalides and as Alkaline Earths in Iodide Salts," *Science* **307**(5707), 231–235 (2005).

27. G. L. Gutsev, "FeO$_4$: A Unique Example of a Closed-Shell Cluster Mimicking a Superhalogen," *Phys. Rev. A* **59**, 3681–3684 (1999).

28. S. C. Rasmussen, "On the Origin of 'Synthetic Metals'," *Mat. Today* **19**(5), 244–245 (2019).

29. M. M. Labes, P. Love, and L. F. Nichols, "A Metallic, Superconducting Polymer," *Chem. Rev.* **79**(1), 1–15 (1979).

30. G. Restrepo, "Challenges for the Periodic Systems of Elements: Chemical, Historical and Mathematical Perspectives," *Chem. Eur. J.* **25**, 15430–15440 (2019).

31. R. Ababei *et al.*, "Making Practical Use of the *pseudo*-Element Concept: An Efficient Way to Ternary Intermetalloid Clusters by an Isoelectronic Pb$^-$Bi Combination," *Chem. Commun.* **48**, 11295–11297 (2012).

Index